SOME EXERCISES
IN PURE MATHEMATICS

WITH EXPOSITORY COMMENTS

J. D. WESTON

Professor of Pure Mathematics
University College of Swansea

AND

H. J. GODWIN

Reader in the Department of Pure
Mathematics, University
College of Swansea

CAMBRIDGE
AT THE UNIVERSITY PRESS
1968

CAMBRIDGE UNIVERSITY PRESS
Cambridge, New York, Melbourne, Madrid, Cape Town, Singapore, São Paulo, Delhi

Cambridge University Press
The Edinburgh Building, Cambridge CB2 8RU, UK

Published in the United States of America by Cambridge University Press, New York

www.cambridge.org
Information on this title: www.cambridge.org/9780521095617

First published 1968
Re-issued in this digitally printed version 2008

A catalogue record for this publication is available from the British Library

Library of Congress Catalogue Card Number: 68-26989

ISBN 978-0-521-09561-7 paperback

CONTENTS

PREFACE

In mathematics, 'advanced' ideas are apt to become 'elementary' with the passage of time. Progress of this kind is not always genuine: mathematical teaching (in universities as well as elsewhere) has often debased itself by superficial treatment of matters that cannot be properly understood without deep and careful thought. But in recent years great changes (representing, for the most part, genuine progress) have taken place in university mathematics, and corresponding changes at other levels are now urgently needed. Our original purpose in compiling this book was to offer some practical help to those who are trying to narrow the gap (or, if they are students, to bridge the gap) between 'school mathematics' and 'university mathematics' as these terms are now generally understood in England and Wales. In countries where the standard period of undergraduate study is rather short, it is necessary for educational efficiency, as well as being desirable on other grounds, that a student entering a university to study mathematical subjects should have had a sound and fairly up-to-date introduction to *pure* mathematics: he should have acquired some understanding of the logical structure and the conceptual basis of this subject, and a habit of careful and critical thinking, as well as a modicum of technical skill of the traditional kind. He cannot be expected to do this without some rigorous training; but if he has some talent for the subject he is likely to enjoy such training, and to benefit from it even if he does not become a mathematical specialist.

In practice, pre-university training in a fundamental subject has to culminate in an official examination based on a published syllabus. Accepting this constraint, and recognizing that proper teaching of the subject was scarcely possible within the framework of a standard 'Advanced-level' syllabus, a small team of teachers and university mathematicians in Swansea undertook, in 1963, to produce and test a radically new syllabus in pure mathematics. The syllabus in its present form was adopted in 1964 as a basis for sixth-form studies in four schools, and the first official examinations based on it were held in 1966. It has also been the basis of some special university courses given in Swansea during the last few years; and it was the original basis of this book.

A syllabus can be very misleading if it is regarded as a mere list of topics. This danger is particularly great for elementary pure mathematics, in which faulty presentation of important ideas has become so well established that it is often regarded as inevitable, and even desirable. For some of the topics in the new syllabus the manner of treatment is all-important; and while we are well aware that the occasional 'deliberate mistake' can be very instructive, we believe that these topics can and should be presented in a way which is logically sound without being forbiddingly difficult. To give some guidance to teachers and pupils working to the syllabus we prepared collections of exercises, with solutions and comments, paying particular attention to those topics on which suitable exercises were not easily available from existing sources. It is this material, with substantial additions, that we now make more widely available, through the enterprise of the Cambridge University Press. We are thus providing a skeletal textbook for the less traditional parts of the new syllabus; we believe that a competent teacher will have no great difficulty in putting suitable flesh on these bones, and that we have left him as much freedom as is practicable to do this in his own individual way. Moreover, although the book was originally planned with the needs of teachers principally in mind, we think it will be of use to students who are taking, or are preparing to take, first-year university courses in pure mathematics.

The arrangement is as follows. The exercises are grouped into four 'collections', each of which ranges over the whole of the new syllabus. The first three collections comprise 140 exercises which vary considerably in depth and in difficulty; none of these is meant to be regarded as a typical examination question, and several of them (especially in the third collection) are designed to introduce the student to interesting topics which lie outside, though not far outside, the scope of the syllabus. The fourth collection consists of the 60 questions that made up the official examination papers of 1966 and 1967 (papers which had the same form, though not the same content, as those of the examinations held concurrently on the 'standard' syllabus to which the new syllabus is alternative). Following the exercises there are detailed solutions, with expository comments, for those in the first collection, and then we give 'hints and comments', with fairly detailed explanations in some cases, for those in the second collection (which is similar in character to the first). Some of the comments contain suggestions for further exercises. An appendix, the first of three, contains the new syllabus with some attendant notes; one of these notes concerns the treatment of elementary analysis, and on this delicate but vitally important matter we have included a further note which constitutes the second appendix. In the third appendix we define a few technical terms, supplementing the numerous definitions, and indications of usage, that appear in other parts of the book (as shown by the index).

We are indebted to the Welsh Joint Education Committee for permission to reproduce the examination papers, and to friends and colleagues who have given us helpful criticisms. We wish to thank particularly Mrs M. M. Williams, of the Amman Valley Grammar School, Carmarthenshire, for her enthusiastic but not uncritical support, and for detecting a number of errors in early drafts.

J. D. W.
H. J. G.

University College of Swansea
October 1967

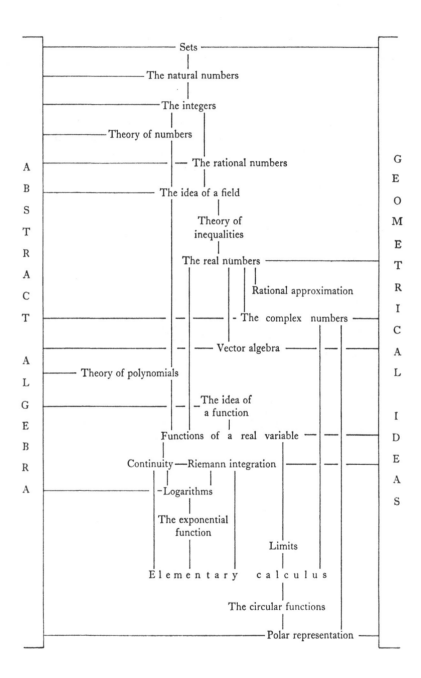

FIRST COLLECTION OF
EXERCISES (1–50)

1. Assuming that A and B are subsets of some given set, explain in words the information conveyed by each of the following equations:

(i) $A \cup B = A$;　　(ii) $A \cap B = B$;　　(iii) $A \setminus B = A$;

(iv) $A \setminus B = B$;　　(v) $A \cap B = \varnothing$;　　(vi) $A \setminus B = \varnothing$.

2. Show that if A, B, and C are non-empty subsets of a given set then

$$A \cup B = B \cup A,$$
$$A \cup (B \cup C) = (A \cup B) \cup C.$$

Show also that if A has only one element and is disjoint from $B \cup C$, and if there is a one-to-one correspondence between $A \cup B$ and $A \cup C$, then there is a one-to-one correspondence between B and C.

What connexions are there between these results and the laws of addition for natural numbers?

3. Explain *the principle of mathematical induction*. Use it to prove that (a) if $E_1, \ldots, E_n$ are finite subsets of a given set then the set

$$E_1 \cup E_2 \cup \ldots \cup E_n$$

is finite, and (b) if n is any natural number then

$$1^2 + 3^2 + 5^2 + \ldots + (2n-1)^2 = \tfrac{1}{3}(4n^3 - n).$$

4. Prove that there are infinitely many prime numbers of the form $3k + 2$, and infinitely many of the form $4k + 3$, k being an integer in each case.

5. Show that, for any positive integer n, the least common multiple of the numbers $1, 2, 3, \ldots, n$ can be expressed as $2^m k$, where k is an odd number and m is the greatest integer such that $2^m \leqslant n$. Hence, or otherwise, prove that, if $n > 1$,

$$1 + \frac{1}{2} + \frac{1}{3} + \ldots + \frac{1}{n}$$

can be expressed as r/s, where r is odd and s is even.

1

Prove similarly that if $n > 1$ then

$$1 + \frac{1}{3} + \frac{1}{5} + \ldots + \frac{1}{2n-1}$$

is not an integer.

6. Prove that if m and n are integers with no common factor, there are integers r and s such that

$$\frac{1}{mn} = \frac{r}{m} + \frac{s}{n}.$$

Deduce from this, or prove otherwise, that every rational number can be expressed in the form

$$\frac{a_1}{b_1} + \ldots + \frac{a_n}{b_n},$$

where the numerators $a_1, \ldots, a_n$ and the denominators $b_1, \ldots, b_n$ are integers and each denominator is a prime number or a power of a prime number.

7. Explain what is meant by *the ring of integers modulo m*, and prove that this ring is a field if and only if m is prime.

Find the reciprocal of 5 in the ring of integers modulo 7, and determine whether or not 5 has a reciprocal in the ring of integers modulo 9.

8. If A and B are subsets of a given set E, the set $A \triangle B$ (the *symmetric difference* of A and B) is defined as follows:

$$A \triangle B = (A \setminus B) \cup (B \setminus A).$$

Show that the class of all subsets of E is an Abelian group with respect to the operation $\triangle$. Show also that the class forms a commutative ring if $\triangle$ is regarded as addition and intersection is regarded as multiplication. Show that this ring has a unit element, and determine the elements that have reciprocals in the ring.

9. Define the term *field*. Prove that, in any field,

$$(-x)(-y) = xy$$

and, if $x \neq 0$,

$$(-x)^{-1} = -x^{-1}.$$

Prove also that if, in a field, $xy = xz$ then either $x = 0$ or $y = z$.

10. Prove that no element of a field has more than two square roots in the field.

Prove that 2 has no square root in the field of rational numbers. Determine whether or not it has a square root in the field of integers modulo 7.

11. Let F be a set of rational numbers forming a field with respect to the usual operations of addition and multiplication. Prove that F consists of all the rational numbers.

12. Explain what is meant by a *totally ordered field*. Show that any totally ordered field contains the system of rational numbers. What special property distinguishes the system of real numbers from that of the rational numbers? How do we know that every positive real number has a real square root?

Give an example of a totally ordered field which is neither the system of rational numbers nor the system of real numbers.

13. What is meant by the term *upper bound* in relation to a set of real numbers? If E is a set that has upper bounds, what is the number $\sup E$?

Suppose that E is the set of all rational numbers that are less than a certain real number x. Assuming that between any two real numbers there is a rational number, prove that $\sup E = x$.

14. Let c be a positive real number, and let E be the set of all numbers x such that $x^3 \leqslant c$. Prove that $c+1$ is an upper bound of E, and that if $b = \sup E$ then $b^3 = c$. Hence show that every real number has a real cube root.

15. If E and H are sets of real numbers, the *product set* EH consists of all numbers that can be expressed in the form xy with x in E and y in H. Prove that if E and H are bounded sets of positive numbers then EH is bounded and $$\sup EH = \sup E \sup H.$$

Show that there are bounded sets E and H for which this equation does not hold.

16. A certain real number x is estimated as $1\cdot022$, and it is known that this estimate is accurate to 3 decimal places (4 significant figures). Determine whether or not we can be sure, from this information alone, (i) that $1\cdot044$ is an accurate estimate of x^2 to 3 decimal places, (ii) that $1\cdot011$ is an accurate estimate of $\sqrt{x}$ (the positive square root of x) to 3 decimal places.

17. Let k be a number greater than 1, and let E be a set of positive real numbers. Prove that $\inf E > 0$ if and only if there is a number x in E such that $x < k \inf E$.

Let E_1 be the set of all numbers k^{-n}, where $n = 1, 2, 3, \ldots$, and let E_2 be the set of all numbers $k^n/n!$. Prove that $\inf E_1 = 0$ and that $\inf E_2 = 0$.

18. Prove that if a, b, c are real numbers such that

$$ax^2 + 2bx + c \geqslant 0$$

for every real number x, then $b^2 \leqslant ac$. Deduce, or prove otherwise, that if $a_1, ..., a_n$ and $b_1, ..., b_n$ are real numbers then

$$(a_1 b_1 + ... + a_n b_n)^2 \leqslant (a_1{}^2 + ... + a_n{}^2)(b_1{}^2 + ... + b_n{}^2)$$

and

$$\{(a_1 + b_1)^2 + ... + (a_n + b_n)^2\}^{\frac{1}{2}} \leqslant (a_1{}^2 + ... + a_n{}^2)^{\frac{1}{2}} + (b_1{}^2 + ... + b_n{}^2)^{\frac{1}{2}}.$$

19. What is a *complex number*, and how are the sum and the product of two complex numbers defined?

Assuming that in the field of real numbers every positive number has a square root, prove that in the field of complex numbers every number has a square root. Deduce that every quadratic equation with complex coefficients has a root in the field of complex numbers.

20. Suppose that F is a field having the property that, for any x in F,

$$x^2 + 1 \neq 0.$$

Let addition and multiplication be defined for the set $F \times F$ as follows:

$$(x_1, y_1) + (x_2, y_2) = (x_1 + x_2, y_1 + y_2),$$
$$(x_1, y_1)(x_2, y_2) = (x_1 x_2 - y_1 y_2, x_1 y_2 + x_2 y_1).$$

Show that $F \times F$, with these operations, is a field, having a subfield which can be identified in a natural way with F, and that in this field the equation

$$z^2 + 1 = 0$$

has a solution.

Determine whether or not it is possible for F to be a finite field.

21. What is the *conjugate*, α^*, of a complex number α? Prove that if α and β are any complex numbers then

$$(\alpha + \beta)^* = \alpha^* + \beta^* \quad \text{and} \quad (\alpha\beta)^* = \alpha^*\beta^*.$$

Suppose that a_0, a_1, ..., a_n are real numbers, and that the complex number λ is a root of the equation

$$a_0 x^n + a_1 x^{n-1} + ... + a_{n-1} x + a_n = 0.$$

Prove that λ^* is also a root of this equation.

22. What is a 3-*dimensional real vector*, and how are linear combinations of such vectors defined? Show that if E is a set consisting of all the linear combinations of certain specified vectors, then every linear combination of elements of E is itself an element of E.

Suppose that E consists of all the linear combinations of two given vectors. Prove that if $\mathbf{p}$, $\mathbf{q}$, and $\mathbf{r}$ are elements of E then one of these three vectors is a linear combination of the other two.

23. How is the *scalar product* of two 3-dimensional real vectors defined? What is meant by the *length* of a vector, and by the statement that two given vectors are *orthogonal* to each other?

If $\mathbf{p} = (1, 0, 2)$ and $\mathbf{q} = (2, 2, -1)$, verify that $\mathbf{p}$ and $\mathbf{q}$ are orthogonal to each other, and find a vector which has the same length as $\mathbf{p}$ and is orthogonal to both $\mathbf{p}$ and $\mathbf{q}$.

24. If $\mathbf{r}_1$ and $\mathbf{r}_2$ are 3-dimensional real vectors, how is the *vector product* $\mathbf{r}_1 \wedge \mathbf{r}_2$ defined? Prove that

$$\mathbf{r}_1 \wedge \mathbf{r}_2 = -\mathbf{r}_2 \wedge \mathbf{r}_1,$$

and that $\mathbf{r}_1 \wedge \mathbf{r}_2$ is orthogonal to $\mathbf{r}_1$ and to $\mathbf{r}_2$.

If $\mathbf{r}_1, \mathbf{r}_2, \mathbf{r}_3$ are vectors such that $\mathbf{r}_1 . \mathbf{r}_2 \wedge \mathbf{r}_3 = 1$, and if

$$\mathbf{r} = \alpha_1 \mathbf{r}_1 + \alpha_2 \mathbf{r}_2 + \alpha_3 \mathbf{r}_3,$$

prove that $\qquad\qquad \alpha_1 = \mathbf{r} . \mathbf{r}_2 \wedge \mathbf{r}_3,$

and obtain similar expressions for the scalars α_2 and α_3.

25. Show that any linear combination of two vectors $\mathbf{u}$ and $\mathbf{v}$ is orthogonal to the vector product $\mathbf{u} \wedge \mathbf{v}$. Deduce that if $\mathbf{u} \wedge \mathbf{v} \neq \mathbf{0}$ then the vectors $\mathbf{u}$, $\mathbf{v}$, and $\mathbf{u} \wedge \mathbf{v}$ are linearly independent.

Assuming that no 4 vectors can be linearly independent, prove that if $\mathbf{u} \wedge \mathbf{v} \neq \mathbf{0}$ then any vector orthogonal to $\mathbf{u} \wedge \mathbf{v}$ is a linear combination of $\mathbf{u}$ and $\mathbf{v}$.

26. Let $\mathbf{e}$ be the vector $(1, 1, 1)$ and let C be the set of all vectors $\mathbf{r}$ that satisfy both of the following conditions:

$$\mathbf{r}^2 = 1, \quad \mathbf{r} . \mathbf{e} = 1.$$

Show how C can be put into one-to-one correspondence with the set of all complex numbers of unit modulus (*the unit circle*).

27. A 4-dimensional real vector (x, y, z, t) may be identified with the ordered pair $(\mathbf{r}, t)$, where $\mathbf{r}$ is the 3-dimensional vector (x, y, z). Such 4-dimensional vectors can be multiplied according to the following definition:

$$(\mathbf{r}_1, t)(\mathbf{r}_2, t_2) = (\mathbf{r}_1 \wedge \mathbf{r}_2 + t_1 \mathbf{r}_2 + t_2 \mathbf{r}_1, t_1 t_2 - \mathbf{r}_1 . \mathbf{r}_2).$$

Show that this multiplication satisfies the associative law but not the commutative law. (The identity

$$(\mathbf{r}_1 \wedge \mathbf{r}_2) \wedge \mathbf{r}_3 - \mathbf{r}_1 \wedge (\mathbf{r}_2 \wedge \mathbf{r}_3) = \mathbf{r}_1 . \mathbf{r}_2 \mathbf{r}_3 - \mathbf{r}_2 . \mathbf{r}_3 \mathbf{r}_1$$

may be assumed.)

Show that, for any $\mathbf{r}$ and any t,

$$(0, 1)(\mathbf{r}, t) = (\mathbf{r}, t)(0, 1) = (\mathbf{r}, t),$$

and calculate the product $(\mathbf{r}, t)(-\mathbf{r}, t)$. Show that if $(\mathbf{r}, t) \neq (0, 0)$ there is a 4-dimensional real vector $(\mathbf{s}, u)$ such that

$$(\mathbf{s}, u)(\mathbf{r}, t) = (\mathbf{r}, t)(\mathbf{s}, u) = (0, 1).$$

28. Find the highest common factor of the polynomials $x^4 + x^2 + 1$ and $x^4 + 2x^3 + x^2 - 1$.

29. Let
$$f(x) = a_0 x^n + a_1 x^{n-1} + \dots + a_{n-1} x + a_n,$$

where the coefficients $a_0, a_1, \dots, a_n$ are integers. Prove that if p is an integer and q is a positive integer, and if $f(p/q) \neq 0$, then

$$|f(p/q)| \geqslant 1/q^n.$$

Use the remainder theorem to deduce that if λ is an irrational number such that $f(\lambda) = 0$ then there is a positive number k, independent of p and q, such that p/q differs from λ by at least k/q^n.

30. If f is a function whose range lies in a given set Y, and if $E \subseteq Y$, how is the *counter-image* $f^{-1}(E)$ defined? Prove that, if $E \subseteq Y$ and $H \subseteq Y$,

$$f^{-1}(E \cup H) = f^{-1}(E) \cup f^{-1}(H),$$

$$f^{-1}(E \cap H) = f^{-1}(E) \cap f^{-1}(H).$$

Taking Y to be the set of all positive integers, give an example of a function f with the property that whenever E consists of one point the set $f^{-1}(E)$ consists of two points.

31. Let f be a function mapping a set X into a set Y, and let φ be a function mapping Y into X. How are the *composite functions* $f \circ \varphi$ and $\varphi \circ f$ defined? If $\varnothing$ is the empty subset of X, what is $(\varphi \circ f)(\varnothing)$?
Show that if $E \subseteq Y$ then $(f \circ \varphi)^{-1}(E) = \varphi^{-1}(f^{-1}(E))$.

32. Let f be a real-valued function whose domain includes all the positive rational numbers, and suppose that if s and t are any such numbers (possibly equal) then

$$f(st) = f(s) + f(t).$$

Prove that $f(1) = 0$.

Assuming that 1 is the only value of s for which $f(s) = 0$, prove also that if s and t are distinct positive rational numbers then $f(s) \neq f(t)$. With the same assumption, deduce that there is no positive rational number s such that $2f(s) = f(2)$.

33. Let f be the function defined as follows:

$$f(x) = \begin{cases} x \text{ if } x \text{ is a rational number,} \\ -x \text{ if } x \text{ is an irrational number.} \end{cases}$$

Prove that (i) f is not monotonic on any interval, (ii) f is inverse to itself.

34. Let f be the function defined by the equation

$$f(x) = x^5 + x^3,$$

where x is any real number. By appealing to a fundamental theorem on continuous functions, prove that if y is any real number then $y = f(x)$ for some value of x.

Prove that f has an inverse function, f^{-1}, and that this cannot be represented by a polynomial. Determine whether or not f^{-1} is a continuous function.

35. Prove that the equation
$$x^3 + x - 1 = 0$$

has exactly one real root, and that this lies between $0 \cdot 68$ and $0 \cdot 69$.

36. Suppose that f is a continuous increasing function on an interval I, and that a systematic method is available for computing $f(x)$ to any degree of accuracy when x is any rational number in I. If y is a given number in the interval $f(I)$, and if ϵ is a given positive number, show how it may be possible to find, by trial, rational numbers x_1 and x_2 in I, with $|x_1 - x_2| < \epsilon$, for which we can be certain that

$$f(x_1) < y < f(x_2).$$

What numerical information do we then have about $f^{-1}(y)$?

37. A certain function f, of a positive real variable, is known to be continuous and increasing. Also, $f(3 \cdot 16)$ and $f(3 \cdot 17)$ are given, to an accuracy of 3 decimal places, by $4 \cdot 997$ and $5 \cdot 011$ respectively. Show that

$$3 \cdot 16 < f^{-1}(5) < 3 \cdot 17.$$

Show also that these inequalities could not have been validly inferred if $f(3 \cdot 16)$ and $f(3 \cdot 17)$ had been specified to an accuracy of only 2 decimal places.

38. Let f be a bounded function on the interval $[0, 1]$, and when $0 \leqslant x \leqslant 1$ let $\qquad \varphi(x) = \sup\{f(x') : 0 \leqslant x' \leqslant x\}.$

Prove that φ is a non-decreasing function on $[0, 1]$, and that φ is continuous if f is continuous.

39. Sketch the graph of the function f defined as follows:

$$f(x) = \begin{cases} x^{\frac{1}{4}} & \text{if } x \geqslant 0, \\ (-x)^{\frac{1}{4}} & \text{if } x < 0. \end{cases}$$

Using theorems on continuity, prove that f is continuous.

40. For any real number x, $[x]$ is defined to be the greatest integer which is not greater than x. Sketch the graph of $[x^2]$ for $-1 \leqslant x \leqslant 2$.

Evaluate
$$\int_1^2 [x^2]\, dx.$$

41. Prove, by induction or otherwise, that for every positive integer n,

$$1^2 + 2^2 + \ldots + n^2 = \tfrac{1}{6}n(n+1)(2n+1).$$

If f is the function defined on the interval $[0, 1]$ by the equation

$$f(x) = x^2,$$

calculate the upper and lower Riemann sums of f corresponding to a division of the interval into n subintervals of equal length. Hence evaluate
$$\int_0^1 x^2\, dx.$$

42. Prove that if $0 \leqslant x \leqslant \tfrac{1}{4}$ then

$$1 + \tfrac{1}{2}x \leqslant \frac{1}{\sqrt{(1-x)}} \leqslant 1 + \tfrac{2}{3}x.$$

Hence show that $\qquad \dfrac{25}{48} \leqslant \displaystyle\int_0^{\frac{1}{2}} \frac{dx}{\sqrt{(1-x^2)}} \leqslant \dfrac{19}{36}.$

43. Define $\log x$, where x is a positive number. Prove that if $t \geqslant 0$ and n is any positive integer, the number

$$t - \tfrac{1}{2}t^2 + \tfrac{1}{3}t^3 - \ldots + \frac{1}{2n+1}t^{2n+1}$$

exceeds $\log(1+t)$ by a number which is not greater than $t^{2n+2}/(2n+2)$.

Prove that $\tfrac{43}{192}$ exceeds $\log\tfrac{5}{4}$ by less than 10^{-3}.

44. Show, without appealing directly to a definition of continuity, that each of the following expressions represents a continuous function of the real variable x:

(i) $x^4 + x^3 + x^2$,   (ii) $x/(1 + x^2)$,

(iii) $x \log(1 + x^2)$,   (iv) e^{-x^2}.

Determine in each case whether the function is bounded or unbounded.

45. How is *the exponential function* of a real variable defined, and how do we know that $(d/dx) e^x = e^x$ for every real number x?

Let a, b, c be three distinct real numbers, and suppose that α, β, γ are real numbers such that, for every real number x,

$$\alpha e^{ax} + \beta e^{bx} + \gamma e^{cx} = 0.$$

Prove that $\alpha = \beta = \gamma = 0$.

46. Let f be a real-valued function which is differentiable throughout a bounded closed interval $[a, b]$. How do we know that there is a point x_0 in $[a, b]$ such that $f(x_0) \geqslant f(x)$ for every x in $[a, b]$? Prove that if $x_0 > a$ then $f'(x_0) \geqslant 0$, and that if $x_0 < b$ then $f'(x_0) \leqslant 0$.

47. Suppose that a function φ, defined on an interval I, has a local maximum or a local minimum at a certain point x of I. Prove that if φ is differentiable at x, and if x is not an end-point of I, then $\varphi'(x) = 0$.

Let f and g be real-valued continuous functions on a bounded closed interval $[a, b]$, where $a < b$, and suppose that $f(a) = g(a)$ and $f(b) = g(b)$. Prove that if f and g are differentiable at every point x such that $a < x < b$, then $f'(x) = g'(x)$ for at least one such point.

48. Let f be a real-valued function defined on an interval I, and suppose that f is differentiable at every point of I. Let a and b be points of I such that $a < b$ and $f'(a) \neq f'(b)$, and let η be a number between $f'(a)$ and $f'(b)$. Let the function g be defined on the closed interval $[a, b]$ as follows:

$$g(x) = f(x) - \eta x \quad (a \leqslant x \leqslant b).$$

Show that g is differentiable at every point of $[a, b]$, and that one of the numbers $g'(a)$, $g'(b)$ is positive and the other negative. Hence show that there is a point ξ of the open interval $]a, b[$ such that $g(\xi)$ is either $\sup g([a, b])$ or $\inf g([a, b])$. Deduce that $f'(\xi) = \eta$, and hence that the set $f'(I)$ is an interval.

49. Suppose that f is a real-valued continuous function on a closed interval $[a, b]$, where $a < b$, and suppose that f is differentiable at every point between a and b. What information does the mean-value theorem give concerning the ratio $\{f(b)-f(a)\}/(b-a)$?

Assuming only that

   (i) $\sin\theta$ increases continuously from 0 to 1 as θ increases from 0 to $\frac{1}{2}\pi$,

   (ii) $\dfrac{\mathrm{d}}{\mathrm{d}\theta}\sin\theta = \cos\theta$,

   (iii) $\sin^2\theta + \cos^2\theta = 1$,

use the mean-value theorem to prove that if $0 < \theta < \frac{1}{2}\pi$ then

$$\cos\theta < \frac{\sin\theta}{\theta} < 1.$$

50. Express $$\frac{x^3+4x^2-10}{(x+3)^2(x+2)^3}$$

in partial fractions, and evaluate

$$\int_{-1}^{1} \frac{x^3+4x^2-10}{(x+3)^2(x+2)^3}\,\mathrm{d}x.$$

SECOND COLLECTION OF
EXERCISES (51–100)

51. Let A, B, and C be subsets of a given set, and let
$$D = B\backslash A, \quad E = C\backslash(A \cup B).$$
Show that the sets A, D, and E are disjoint from one another, and that
$$A \cup B \cup C = A \cup D \cup E.$$

In the following table each numerical entry gives the number of elements common to the sets whose names appear at the top of its column and to the left of its row. The number of elements of $A \cap B \cap C$ is 9. Find the number of elements of $A \cup B \cup C$.

	A	B	C
A	43		
B	16	28	
C	14	13	27

52. Show that if A, B, and C are non-empty subsets of a given set then there is a one-to-one correspondence between the Cartesian products $A \times B$ and $B \times A$, and that there is a one-to-one correspondence between $A \times (B \times C)$ and $(A \times B) \times C$. Show also that if the sets are finite and there is a one-to-one correspondence between $A \times B$ and $A \times C$ then there is a one-to-one correspondence between B and C.

What connexions are there between these results and the laws of multiplication for natural numbers?

53. Let Q be a set of natural numbers, and for natural numbers m and n let the statement
$$m \sim n$$
mean that there is a number q belonging to Q such that $m = nq$ or $n = mq$. Show that the binary relation $\sim$ so defined is an equivalence relation if and only if, for some number h, Q consists of all the numbers
$$1, \quad h, \quad h^2, \quad h^3, \quad \ldots.$$

54. Make a list of all the prime numbers between 1 and 100.
Factorize the number 102427689.

11

55. Prove that if $p_1, ..., p_n$ are distinct prime numbers then the sum

$$\frac{1}{p_1} + ... + \frac{1}{p_n}$$

is not an integer. Find the least value of n for which this sum can be greater than 1.

56. Define the terms *binary operation, commutative law, associative law*.

Give an example of (i) a binary operation which is commutative but not associative, (ii) a binary operation which is associative but not commutative.

57. Define the terms *group, Abelian group, ring*.

Prove that in any ring R there is an element p such that $px = xp = p$ for every x in R. Prove that there is only one such element p in R. Decide whether or not it is possible for R to have more than one element q such that $qx = xq = p$ for every x in R.

58. Let R_1 and R_2 be rings, each having more than one element, and let addition and multiplication be defined for the set $R_1 \times R_2$ as follows:

$$(x_1, x_2) + (y_1, y_2) = (x_1 + y_1, x_2 + y_2),$$
$$(x_1, x_2)(y_1, y_2) = (x_1 y_1, x_2 y_2).$$

Show that $R_1 \times R_2$, with these operations, is a ring which cannot be isomorphically embedded in a field. Determine all the elements that have reciprocals in this ring.

59. Let R be a commutative ring with unit element, and suppose that R has an element u, not 0 or 1, such that $u^2 = u$. Let

$$R_1 = \{ux : x \in R\}.$$

Prove that R_1 is a subring of R, and that R_1 has a unit element. Prove also that there is another subring R_2 of R such that (i) R_2 has a unit element, (ii) every element of R can be expressed in the form $x_1 + x_2$, where $x_1 \in R_1$ and $x_2 \in R_2$, (iii) if $x_1 \in R_1$ and $x_2 \in R_2$ then $x_1 x_2 = 0$, (iv) $R_1 \cap R_2$ consists of the single element 0. Show that the correspondence

$$x_1 + x_2 \sim (x_1, x_2)$$

is one-to-one between R and $R_1 \times R_2$.

60. Give an example of a field in which every element has exactly one square root. Prove that there is no finite field in which every non-zero element has two square roots.

61. Prove that if a field F has a non-zero element a such that $a + a = 0$, then $x + x = 0$ for every element x of F.

Let G be the multiplicative group formed by the non-zero elements of a field F, and suppose that G is isomorphic to a subgroup of the additive group of F. Prove that F has only two elements.

62. Prove that the following statements about a totally ordered field F are equivalent (in the sense that if any one of them is true then the others are true):

  (i) every non-empty subset of F which has upper bounds in F has a least upper bound in F;

  (ii) every non-empty subset of F which has lower bounds in F has a greatest lower bound in F;

  (iii) if L and R are non-empty subsets of F such that $F = L \cup R$ and every element of L is less than every element of R, then either L has a greatest element or R has a least element.

63. Prove that if p and q are positive rational numbers and $\sqrt{p}$ is irrational then $\sqrt{p} + \sqrt{q}$ is irrational.

64. Let n be an integer greater than 1. Prove that if x is a real number greater than 1 then
$$n(x-1) < x^n - 1 < nx^{n-1}(x-1).$$
Hence show that if $a > b > 0$ then
$$nb^{n-1}(a-b) < a^n - b^n < na^{n-1}(a-b),$$
and deduce from one of these inequalities that no positive real number has more than one positive nth root.

65. Let E be a non-empty set of real numbers such that if $x \in E$ and $y \in E$ then $x - y \in E$. Prove that E is a group with respect to addition.

Suppose further that there are infinitely many points of E between 0 and 1. Prove that, for every positive integer k, at least one interval of length $1/k$ contains more than one point of E, and deduce that every non-degenerate interval contains a point of E.

Prove that if λ is an irrational number then between any two real numbers there is a number of the form $m + n\lambda$, where m and n are integers.

66. Let c be a positive real number, and let E be the set of all numbers that are squares of rational numbers and are less than c. Prove that $c = \sup E$.

67. Prove that the following statements about a totally ordered field F are equivalent:

  (i) the set of integers is unbounded in F;
  (ii) if $0 < x < y$ in F then $nx > y$ for some integer n;
  (iii) 0 is the greatest lower bound in F of the set of reciprocals of positive integers;
  (iv) between any two elements of F there is a rational number;
  (v) each element x of F is the least upper bound in F of the set of rational numbers that are less than x.

68. For any positive rational number p, let F_p be the set of all numbers of the form $r + s\sqrt{p}$, where r and s are rational numbers. Prove that F_p is a subfield of the field of real numbers, and is the smallest subfield containing $\sqrt{p}$.

Prove that if p and q are distinct prime numbers then the fields F_p and F_q are not isomorphic to each other.

69. Complex numbers a, b, c are given, and $a \neq 0$. It is required to assign to each natural number n a complex number u_n in such a way that, for every n,
$$au_{n+2} + 2bu_{n+1} + cu_n = 0.$$
Show that this 'recurrence relation' allows u_1 and u_2 to be chosen arbitrarily and that u_n is then uniquely determined for every n.

Show that the recurrence relation is satisfied if
$$u_n = \alpha\lambda^n + \beta\mu^n$$
for every n, where λ and μ are the roots of the quadratic equation
$$ax^2 + 2bx + c = 0$$
and where α and β are any complex numbers independent of n. Show further that if $b^2 \neq ac$ then α and β can be calculated in terms of u_1 and u_2 when these have been chosen arbitrarily.

Show that if a, b, c, u_1, u_2 all belong to a certain subfield F of the field of complex numbers then $u_n \in F$ for every n although λ and μ may not belong to F.

70. If z_1 and z_2 are complex numbers, the *straight segment* joining z_1 and z_2 is, by definition, the set of all complex numbers that can be expressed in the form $\lambda z_1 + (1 - \lambda) z_2$ with $0 \leqslant \lambda \leqslant 1$. A non-empty set C of complex numbers is said to be *convex* if, for any points z_1 and z_2 of C, the straight segment joining z_1 and z_2 is a subset of C. Prove that the intersection of any family of convex sets is either convex or empty.

Let C_1 be the set of all complex numbers z for which $|z| < 3$, and let C_2 be the set for which $\operatorname{re} z + \operatorname{im} z < 2$. Prove that C_1, C_2, and $C_1 \setminus C_2$ are convex, and that $C_2 \setminus C_1$ is not convex.

71. Let there be given an isomorphism, $z \sim z'$, of a field F with itself, and let E be the set of all those elements z of F for which $z' = z$. Prove that E is a subfield of F.

Suppose further that $z'' = z$ for every z in F, and that F has at least one element z such that $z' \neq z$. Prove that F has an element h, not in E, such that $h^2 \in E$, $h' = -h$, and every element z of F can be expressed in the form $x + hy$, where x and y are elements of E uniquely determined by z.

72. Suppose that a, b, and c are real numbers, and that $a \geqslant 0$ and $c \geqslant 0$. Prove that $b^2 < ac$ if and only if

$$a\lambda^2 + 2b\lambda + c > 0$$

for every real number λ.

Prove that if $\mathbf{r}_1$ and $\mathbf{r}_2$ are 3-dimensional real vectors then

$$|\mathbf{r}_1 . \mathbf{r}_2| \leqslant |\mathbf{r}_1|\,|\mathbf{r}_2|,$$

equality holding if and only if there are scalars α_1 and α_2, not both 0, such that $\alpha_1 \mathbf{r}_1 = \alpha_2 \mathbf{r}_2$. Deduce the triangle inequality.

73. Suppose that to each vector $\mathbf{r}$ there corresponds a vector $\mathbf{r}'$ in such a way that $\mathbf{0}' = \mathbf{0}$ and, for any vectors $\mathbf{p}$ and $\mathbf{q}$, the distance between $\mathbf{p}'$ and $\mathbf{q}'$ is the same as that between $\mathbf{p}$ and $\mathbf{q}$. By considering the squared lengths of $\mathbf{p}$, $\mathbf{q}$, and $\mathbf{p} - \mathbf{q}$, prove that

$$\mathbf{p}' . \mathbf{q}' = \mathbf{p} . \mathbf{q}$$

for any vectors $\mathbf{p}$ and $\mathbf{q}$. By considering the squared length of

$$(\alpha\mathbf{p} + \beta\mathbf{q})' - \alpha\mathbf{p}' - \beta\mathbf{q}',$$

where α and β are any scalars, prove also that

$$(\alpha\mathbf{p} + \beta\mathbf{q})' = \alpha\mathbf{p}' + \beta\mathbf{q}'.$$

Prove that the correspondence $\mathbf{r} \sim \mathbf{r}'$ is one-to-one.

74. From a definition of the vector product $\mathbf{r}_1 \wedge \mathbf{r}_2$, for 3-dimensional vectors $\mathbf{r}_1$ and $\mathbf{r}_2$, prove that if α_1 and α_2 are scalars then

$$(\alpha_1 \mathbf{r}_1) \wedge (\alpha_2 \mathbf{r}_2) = \alpha_1 \alpha_2 (\mathbf{r}_1 \wedge \mathbf{r}_2).$$

Assuming the identity

$$(\mathbf{r}_1 \wedge \mathbf{r}_2) \wedge \mathbf{r}_3 = \mathbf{r}_3 . \mathbf{r}_1 \mathbf{r}_2 - \mathbf{r}_2 . \mathbf{r}_3 \mathbf{r}_1,$$

prove that, for any vectors $\mathbf{r}_1$, $\mathbf{r}_2$, $\mathbf{r}_3$,

$$(\mathbf{r}_1 \wedge \mathbf{r}_2) \wedge \mathbf{r}_3 + (\mathbf{r}_2 \wedge \mathbf{r}_3) \wedge \mathbf{r}_1 + (\mathbf{r}_3 \wedge \mathbf{r}_1) \wedge \mathbf{r}_2 = \mathbf{0}.$$

Prove also that $\mathbf{r}_1 \wedge \mathbf{r}_2 = \mathbf{0}$ if and only if there are scalars α_1 and α_2, not both 0, such that $\alpha_1 \mathbf{r}_1 = \alpha_2 \mathbf{r}_2$.

75. Suppose that to each vector $\mathbf{r}$ there corresponds a vector $\mathbf{r}'$ such that $\mathbf{r}.\mathbf{r}' = 0$, and suppose that

$$(\alpha_1 \mathbf{r}_1 + \dots + \alpha_n \mathbf{r}_n)' = \alpha_1 \mathbf{r}_1' + \dots + \alpha_n \mathbf{r}_n'$$

for any vectors $\mathbf{r}_1, \dots, \mathbf{r}_n$ and scalars $\alpha_1, \dots, \alpha_n$. Prove that

$$\mathbf{p}'.\mathbf{q} = -\mathbf{q}'.\mathbf{p}$$

for any vectors $\mathbf{p}$ and $\mathbf{q}$.

Let $\mathbf{i}$, $\mathbf{j}$, $\mathbf{k}$ be a right-handed sequence of mutually orthogonal unit vectors, and let

$$\boldsymbol{\omega} = \mathbf{j}'.\mathbf{k}\,\mathbf{i} + \mathbf{k}'.\mathbf{i}\,\mathbf{j} + \mathbf{i}'.\mathbf{j}\,\mathbf{k}.$$

Prove that $\mathbf{r}' = \boldsymbol{\omega} \wedge \mathbf{r}$ for every vector $\mathbf{r}$.

Prove that if $\mathbf{u}$ is a vector such that

$$\mathbf{r}' = \mathbf{u} \wedge \mathbf{r}$$

for every vector $\mathbf{r}$, then $\mathbf{u} = \boldsymbol{\omega}$.

76. Vectors $\mathbf{u}$ and $\mathbf{v}$ are given, $\mathbf{u} \neq \mathbf{0}$, and λ is a given scalar. It is required to find a vector $\mathbf{r}$ such that

$$\mathbf{u} \wedge \mathbf{r} = \lambda \mathbf{u} - \mathbf{v}.$$

Show that this problem has either no solution or infinitely many solutions, and that it has no solution if $\lambda \neq \mathbf{u}.\mathbf{v}/\mathbf{u}^2$. Using the identity

$$\mathbf{r}_1 \wedge (\mathbf{r}_2 \wedge \mathbf{r}_3) = \mathbf{r}_3.\mathbf{r}_1\mathbf{r}_2 - \mathbf{r}_1.\mathbf{r}_2\mathbf{r}_3,$$

find a solution to the problem on the assumption that $\lambda = \mathbf{u}.\mathbf{v}/\mathbf{u}^2$.

77. A 2-by-2 *matrix*, $\begin{pmatrix} a & b \\ c & d \end{pmatrix}$,

over a given field F, consists of 4 elements a, b, c, d of F, arranged as shown. Such matrices are multiplied according to the following definition:

$$\begin{pmatrix} a_1 & b_1 \\ c_1 & d_1 \end{pmatrix}\begin{pmatrix} a_2 & b_2 \\ c_2 & d_2 \end{pmatrix} = \begin{pmatrix} a_1 a_2 + b_1 c_2 & a_1 b_2 + b_1 d_2 \\ c_1 a_2 + d_1 c_2 & c_1 b_2 + d_1 d_2 \end{pmatrix}.$$

Show that this multiplication obeys the associative law but not the commutative law.

Find 2-by-2 matrices O and I such that, for any 2-by-2 matrix A over F,

$$OA = AO = O,$$
$$IA = AI = A.$$

Also find a 2-by-2 matrix A, other than O, such that

$$A^2 = O.$$

78. State the binomial theorem. Show that in the case of a prime index p all the binomial coefficients except the first and the last are divisible by p.

Prove by mathematical induction that, for every natural number n and every prime number p,

$$n^p \equiv n \pmod{p}.$$

79. Let n be any integer greater than 1. Use the binomial theorem to prove that

$$\left(\frac{n^2}{n^2-1}\right)^n > 1 + \frac{1}{n},$$

and deduce that

$$\left(1-\frac{1}{n}\right)^{-n} > \left(1-\frac{1}{n+1}\right)^{-n-1}.$$

Prove also that

$$\left(1+\frac{1}{n}\right)^n < \left(1-\frac{1}{n}\right)^{-n}.$$

Now let m be any positive integer. Prove that

$$\left(1+\frac{1}{m}\right)^m < \left(1+\frac{1}{m+1}\right)^{m+1},$$

and deduce that

$$\left(1+\frac{1}{m}\right)^m < \left(1-\frac{1}{n}\right)^{-n}.$$

80. For any subset E of a set X, the *characteristic function* of E (relative to X) is the function χ_E defined as follows:

$$\chi_E(x) = \begin{cases} 1 & \text{if } x \in E, \\ 0 & \text{if } x \in X \setminus E. \end{cases}$$

Prove that if $\chi_E = \chi_H$ then $E = H$.

If $E \subseteq X$ and $H \subseteq X$, let the functions $\chi_E + \chi_H$ and $\chi_E \chi_H$ be defined by the equations

$$(\chi_E + \chi_H)(x) = \chi_E(x) \underset{2}{+} \chi_H(x),$$

$$(\chi_E \chi_H)(x) = \chi_E(x)\chi_H(x),$$

where x is any element of X, and where $\underset{2}{+}$ denotes addition modulo 2. Prove that, with these definitions of sum and product, the characteristic functions of all the subsets of X form a ring, in which

$$\chi_E + \chi_H = \chi_{E \triangle H} \quad \text{and} \quad \chi_E \chi_H = \chi_{E \cap H},$$

where $E \triangle H = (E \setminus H) \cup (H \setminus E)$.

81. Let A be an Abelian group, written in the additive notation, and let $\mathscr{A}$ be the set of all functions that map A into itself. Let binary operations $+$ and o be defined on $\mathscr{A}$ as follows: if $f \in \mathscr{A}$ and $g \in \mathscr{A}$ then

$$(f+g)(a) = f(a) + g(a) \quad \text{and} \quad (f \circ g)(a) = f(g(a)),$$

where a is any element of A. Prove that $\mathscr{A}$ is an Abelian group with respect to $+$, and that the operation o is associative.

Let $\mathscr{B}$ be the subset of $\mathscr{A}$ consisting of those functions f which are *additive*, in that if $a \in A$ and $b \in A$,

$$f(a+b) = f(a) + f(b).$$

Prove that $\mathscr{B}$ is a ring with respect to the operations $+$ and o.

Show that if A is the additive group of the integers then $\mathscr{A}$ is not a ring with respect to $+$ and o.

82. Let A be an Abelian group, written in the additive notation, and let f and g be real-valued functions defined on A in such a way that, if $a \in A$ and $b \in A$,

$$f(a-b) = f(a)f(b) + g(a)g(b),$$
$$g(a-b) = g(a)f(b) - f(a)g(b).$$

Prove that $g(0) = 0$, and that if g has more than one value then $f(0) = 1$.

Let χ be the complex-valued function defined on A by the equation

$$\chi(a) = f(a) + ig(a).$$

Assuming that $f(0) = 1$, prove that $|\chi(a)| = 1$ and that

$$\chi(a+b) = \chi(a)\chi(b).$$

Assuming further that the range of f is an interval with mid-point 0, prove that the range of χ is the unit circle.

83. Let f be the function defined as follows:

$$f(x) = \begin{cases} x & \text{if } x \text{ is a rational number,} \\ 0 & \text{if } x \text{ is an irrational number,} \end{cases}$$

and let $\varphi(x) = \sup\{f(x') : x' \leqslant x\}$. Prove that

$$\varphi(x) = \begin{cases} x & \text{if } x \geqslant 0, \\ 0 & \text{if } x < 0. \end{cases}$$

84. Suppose that f and g are real-valued functions defined on the same interval I. Define the function $f+g$. Prove that if f and g are bounded then $f+g$ is bounded and

$$\inf f(I) + \sup g(I) \leqslant \sup (f+g)(I) \leqslant \sup f(I) + \sup g(I).$$

85. Let a and b be real numbers such that $a^3 + b^2 \geqslant 0$, and let α and β be the roots of the quadratic equation

$$x^2 + 2bx - a^3 = 0.$$

Show that α and β are real, and that $\alpha^{\frac{1}{3}} + \beta^{\frac{1}{3}}$ is a root of the cubic equation

$$x^3 + 3ax + 2b = 0.$$

Show that $\{\frac{1}{2} + \frac{1}{2}\sqrt{(31/27)}\}^{\frac{1}{3}} + \{\frac{1}{2} - \frac{1}{2}\sqrt{(31/27)}\}^{\frac{1}{3}}$ is a root of the equation

$$x^3 + x - 1 = 0.$$

86. Let f be a continuous function which maps the set of all real numbers into the interval $[-1, 1]$, and suppose that

$$f(n) = (-1)^n$$

for every integer n. Let g be the function defined as follows:

$$g(x) = \begin{cases} f(1/x) & \text{if } x \neq 0, \\ 0 & \text{if } x = 0. \end{cases}$$

Prove that g is not continuous on any non-degenerate interval containing 0, but that if I is any bounded closed interval then $g(I)$ is a bounded closed interval.

87. A real-valued function f is continuous on a certain interval I, and it is known that for every rational number r in I the number $f(r)$ is rational. It is also known that there are rational numbers a and b in I such that $f(a) > 0$ and $f(b) < 0$. Prove or disprove the assertion that there must be a rational number c in I such that $f(c) = 0$.

88. Let f be a real-valued function which is additive in that

$$f(x+y) = f(x) + f(y)$$

for any real numbers x and y, and let $a = f(1)$. Prove that $f(r) = ar$ for every rational number r, and that if f is continuous then $f(x) = ax$ for every real number x.

89. Sketch the graph of the function f defined as follows:

$$f(x) = \begin{cases} (1-x)^{\frac{1}{2}} & \text{if } x \leqslant 1, \\ (x-1)^2 & \text{if } x > 1. \end{cases}$$

Prove that f is continuous.

90. Let the function f be defined by the equation

$$f(x) = 2^x,$$

where x is any real number. Prove that the range of f is the set of all positive numbers, and that f has a continuous monotonic inverse, f^{-1}. Sketch the graph of f^{-1}.

91. From a definition of the logarithmic function, deduce that (i) if a and b are positive numbers such that $\log a \leqslant \log b$ then $a \leqslant b$, and (ii) if $x > 0$ then $\log x \leqslant x - 1$.

Let $\lambda_1, ..., \lambda_n$ and $x_1, ..., x_n$ be positive numbers such that

$$\lambda_1 x_1 + ... + \lambda_n x_n = \lambda_1 + ... + \lambda_n = 1.$$

Prove that $\qquad\qquad x_1^{\lambda_1} x_2^{\lambda_2} ... x_n^{\lambda_n} \leqslant 1.$

Deduce that if $a_1, ..., a_n, p_1, ..., p_n$ are positive numbers then

$$a_1^{p_1} a_2^{p_2} ... a_n^{p_n} \leqslant \left(\frac{p_1 a_1 + ... + p_n a_n}{p_1 + ... + p_n} \right)^{p_1 + ... + p_n}.$$

92. A real-valued function f is defined on a set E of real numbers which has points in every open interval containing 0. Show that each of the following propositions about f is the negation of the other.

(i) For every positive number ϵ there is a positive number δ such that $|f(x)| < \epsilon$ for every x in E such that $0 < |x| < \delta$.

(ii) There is a positive number ϵ such that for every positive number δ there is an x in E such that $0 < |x| < \delta$ and $|f(x)| \geqslant \epsilon$.

What do these propositions have to do with the theory of limits?

93. Let f be a real-valued function of a real variable, let a and b be real numbers, and suppose that the domain of f intersects every open interval that contains a. What does it mean to say that $f(x) \to b$ as $x \to a$?

Prove that if $f(x) \to 0$ as $x \to 0$, and if α is a positive number, then $|f(x)|^\alpha \to 0$ as $x \to 0$.

94. Prove that $2^x \to 1$ as $x \to 0$. Hence prove that if $a \geqslant b > 0$ and if $f(x) = (a^{1/x} + b^{1/x})^x$ when $x \neq 0$ then

$$f(x) \to a \quad \text{as} \quad x \to 0+,$$
$$f(x) \to b \quad \text{as} \quad x \to 0-.$$

95. The functions sinh and cosh ('hyperbolic sine' and 'hyperbolic cosine') are defined by the equations

$$\sinh x = \tfrac{1}{2}(e^x - e^{-x}), \quad \cosh x = \tfrac{1}{2}(e^x + e^{-x}),$$

where x is any real number. Prove that

$$\sinh(-x) = -\sinh x, \quad \cosh(-x) = \cosh x,$$

$$\frac{d}{dx}\sinh x = \cosh x, \qquad \frac{d}{dx}\cosh x = \sinh x,$$

and, for any real numbers x and y,

$$\sinh(x+y) = \cosh x \sinh y + \sinh x \cosh y,$$

$$\cosh(x+y) = \cosh x \cosh y + \sinh x \sinh y.$$

96. Let the functions f and g be defined by the equations

$$f(x) = |x|, \quad g(x) = x|x|,$$

where x is any real number. Prove that f is not differentiable at 0, and that g is differentiable everywhere. Sketch the graphs of the functions f, g, and g'.

97. Let f and g be functions which have the same domain and are n-times differentiable at a certain point x, for some natural number n. Prove by mathematical induction that the function fg is n-times differentiable at x, and that

$$(fg)^{(n)}(x) = f^{(n)}(x)g(x) + nf^{(n-1)}(x)g'(x) + \ldots + f(x)g^{(n)}(x),$$

where, for $r = 0, 1, 2, \ldots, n$, the coefficient of $f^{(n-r)}(x)g^{(r)}(x)$ in the sum on the right is the *binomial coefficient*

$$\frac{n!}{r!(n-r)!}.$$

98. Consider the function f defined as follows: (i) if x can be expressed in the form k/p^2, where p is a (positive) prime number and k is one of the numbers p, $p+1$, ..., p^2-1, then

$$f(x) = \begin{cases} (k+1)/p^2 & \text{if } k \neq p^2-1, \\ 1/p & \text{if } k = p^2-1; \end{cases}$$

(ii) if x is any other real number then $f(x) = x$. Show that f has an inverse function, f^{-1}, whose domain is that of f. Show also that $f'(0) = 1$ and that f^{-1} is not differentiable at the point 0.

99. Find a function f, whose domain is the set of all positive numbers, such that $f(1) = 1$ and
$$x^2 f'(x) + f(x) = 0$$

for every positive number x. Show that there is only one such function, and that its values are all greater than $1/e$. Show also that this function satisfies the condition
$$f(\tfrac{1}{3}) = \{f(\tfrac{1}{2})\}^2.$$

100. What are the real numbers θ for which $\sin\theta = 0$? Prove that if θ is not one of these numbers then
$$\cos\theta + \cos 3\theta + \ldots + \cos(2n-1)\theta = \frac{\sin 2n\theta}{2\sin\theta}$$
for every positive integer n.

THIRD COLLECTION OF
EXERCISES (101–140)

101. For a given non-empty set X one might be given a rule that assigns to each subset E of X a certain subset $\bar{E}$ of X. Any of the following statements might then be made:
- (1) for every subset H of X, if $E \subseteq H$ then $\bar{E} \subseteq \bar{H}$;
- (2) for every subset H of X, if $E \subset H$ then $\bar{E} \subset \bar{H}$;
- (3) for every subset H of X, if $E \subset H$ then $\bar{E} \subseteq \overline{\bar{H}}$;
- (4) for some subset H of X, if $E \subseteq H$ then $\bar{E} \subset \bar{H}$.

Show that one of these statements is necessarily false, but that each of the other three could be true. For each of the three statements that could be true, determine whether or not it implies either or both of the other two; in particular, determine whether or not there are two logically equivalent statements among these three. Determine also whether or not any of the statements is necessarily true.

102. Let $\mathscr{A}$ be a class of subsets of some given set. If $\mathscr{B} \subseteq \mathscr{A}$ let $\mathscr{B}^\wedge$ be the class of all sets C in $\mathscr{A}$ which are such that if $B \in \mathscr{B}$ then $B \subset C$; and let $\mathscr{B}^\vee$ be defined similarly but with $\supset$ in place of $\subset$. Prove that if $\mathscr{B} \subseteq \mathscr{A}$ and $\mathscr{C} \subseteq \mathscr{A}$ then

$$(\mathscr{B} \cup \mathscr{C})^\wedge = \mathscr{B}^\wedge \cap \mathscr{C}^\wedge,$$

and deduce that if $\mathscr{B} \subseteq \mathscr{C}$ then $\mathscr{B}^\wedge \supseteq \mathscr{C}^\wedge$. Prove also that $\mathscr{B} \subseteq \mathscr{B}^{\wedge\vee}$ and that $\mathscr{B}^{\wedge\vee\wedge} = \mathscr{B}^\wedge$.

103. A *semigroup* is a set with an associative operation $(x, y) \to xy$ (which may be written in the additive notation if it is commutative). An element u of a semigroup is said to be *left-neutral* if $ux = x$ for every element x, *right-neutral* if $xu = x$ for every x, and *neutral* if it is both left-neutral and right-neutral. Prove that if a semigroup has a left-neutral element u and a right-neutral element v, then $u = v$ and no other element is left-neutral or right-neutral. Give an example of a semigroup having more than one left-neutral element, and show that such a semigroup cannot be subject to *the right-cancellation law* (which states that if $xy = zy$ then $x = z$).

Show that any semigroup can be isomorphically embedded in a semi-

23

group that has a neutral element, and that a semigroup can be iso-morphically embedded in an Abelian group if and only if it is subject to the commutative law and the (two-sided) cancellation law.

For any natural number n, the natural numbers that are not less than n form a semigroup with respect to addition, and also with respect to multiplication: show that these two semigroups are not isomorphic to each other.

104. Let G be a group, written in the multiplicative notation. A subset H of G is called a *subgroup* of G if the restriction to $H \times H$ of the mapping $(x, y) \to xy$ is a binary operation with which H is a group. Prove that a non-empty subset H of G is a subgroup of G if and only if $x^{-1}y \in H$ for any elements x, y of H.

Given a subgroup H of G, let '$x \sim y$' mean, for elements x, y of G, that $x^{-1}y \in H$. Verify that $\sim$, thus defined, is an equivalence relation. The corresponding equivalence classes are called the *left cosets* of H in G. Prove that if $x \in G$ then the left coset to which x belongs is $\{xh: h \in H\}$, and show that this coset is in one-to-one correspondence with H. Deduce *Lagrange's theorem*,† which states that the number of elements in a subgroup of a finite group G is a factor of the number of elements in G. Show that the same conclusion could be reached by considering 'right cosets' instead of left cosets.

105. Let x and y be elements of a given commutative ring. Show how, within the ring, $(x^2 + y^2)^2$ can be expressed as the sum of two squares.

106. By definition, a *Boolean ring* is a ring in which $x^2 = x$ for every element x (every element is idempotent). By considering $(x+y)^2$ and $(x-y)^2$, or otherwise, prove that in any Boolean ring multiplication is commutative and addition is the same as subtraction.

Prove that if, in a Boolean ring, u, x, y are such that

$$x = uxy$$

then $x = ux$. Prove also that if x and y are any elements of a Boolean ring, and $u = x + y + xy$, then

$$x = ux \quad \text{and} \quad y = uy.$$

Hence show that for any n elements $x_1, \ldots, x_n$ of the ring there is an element u of the ring such that

$$x_1 = ux_1, \quad x_2 = ux_2, \quad \ldots, \quad x_n = ux_n.$$

† J. L. Lagrange, 1736–1813; French–Italian.

107. Let p be the product of the squares of the first fifty even numbers, and let q be the product of the squares of the first fifty odd numbers Prove that

$$101 < \frac{p}{q} < 200.$$

108. Let R be any ring, and let Z be the ring of integers. By considering $R \times Z$, show that R can be isomorphically embedded in a ring R_1 such that (i) R_1 has a unit element, 1 say, and $1 \notin R$; (ii) if $x \in R$ and $y \in R$ then $xy \in R$ and $yx \in R$. Show that no element of R_1 has a reciprocal in R.

If $x \in R$ and $y \in R$ let

$$x \circ y = x + y - xy.$$

Show that the operation $\circ$, thus defined, is associative, and that $x \circ 0 = 0 \circ x = 0$ for every element x of R. Show also that if $x \in R$, the existence in R of an element x' such that

$$x' \circ x = x \circ x' = 0$$

is necessary and sufficient for $1 - x$ to have a reciprocal in R_1.

109. Construct the addition table and the multiplication table for a field having 4 (and only 4) elements. Verify that the system so represented is a field, and that it is essentially the only field with 4 elements.

110. Prove the following propositions concerning elements x, y of a totally ordered field:

(i) $2xy \leqslant x^2 + y^2$ and $(x+y)^2 \leqslant 2(x^2 + y^2)$;

(ii) $(x+y)(1+x)(1+y) \leqslant 2(1+x^2)(1+y^2)$;

(iii) if $x < y$ and $xy > 0$ then $y^{-1} > x^{-1}$;

(iv) if $x > 1$ and $y > 1$ then $x + y < 2xy$.

111. Prove by induction that if $x > -1$ then

$$(1+x)^n \geqslant 1 + nx$$

for every positive integer n.

If $a_1, \ldots, a_n$ are positive numbers, let

$$b_n = \left(\frac{a_1 + \ldots + a_n}{n}\right)^n, \quad c_n = a_1 a_2 \ldots a_n.$$

Prove that, if $n > 1$, $\qquad \dfrac{b_n}{b_{n-1}} \geqslant a_n,$

and deduce that $\qquad \dfrac{b_n}{c_n} \geqslant \dfrac{b_{n-1}}{c_{n-1}} \geqslant 1.$

WSF

112. An infinite sequence of real numbers a_1, a_2, a_3, ... is such that

$$a_{m+n} \leqslant a_m + a_n$$

for all positive integers m, n. Prove that

$$a_{kn} \leqslant ka_n$$

for all positive integers k, n.

Show that if $m > n$ and k is chosen so that

$$kn < m \leqslant (k+1)n$$

then

$$a_m \leqslant ka_n + a_{m-kn},$$

and hence that if b_n is the greatest of the numbers $a_1, ..., a_n$ then

$$\frac{a_m}{m} \leqslant \frac{kn}{m}\frac{a_n}{n} + \frac{b_n}{m}.$$

Deduce that if ϵ is a given positive number, and n is a given positive integer, then

$$\frac{a_m}{m} \leqslant \frac{a_n}{n} + \epsilon$$

provided that m is greater than a certain number (depending on n).

113. Prove that, for each natural number n,

$$\frac{1}{n+1} + \frac{1}{n+2} + ... + \frac{1}{2n} > \frac{1}{2}$$

and

$$1 + \frac{1}{2} + \frac{1}{3} + ... + \frac{1}{2^n} > \frac{n}{2},$$

where the first sum has n terms and the second has 2^n terms.

For an infinite sequence $u_1, u_2, u_3, ...$ of non-negative numbers, let

$$s_n = u_1 + ... + u_n$$

for each natural number n. Prove that if the set $\{s_n: n = 1, 2, 3, ...\}$ is bounded then

$$\inf\{u_n: n = 1, 2, 3, ...\} = 0.$$

State the converse proposition, and prove that it is false.

114. What is an *interval* of real numbers? Prove that if a non-empty set is an intersection of intervals then it is itself an interval.

Let $\mathscr{C}$ be a class of intervals. Suppose that $\mathscr{C}$ has infinitely many members, and that any two of its members intersect. Prove that, for each natural number n, if $I_1, ..., I_n$ are any n members of $\mathscr{C}$ then

$$I_1 \cap ... \cap I_n \neq \varnothing.$$

Prove or disprove the assertion that there is necessarily a point which belongs to every member of $\mathscr{C}$.

115. Let a, b, c be real numbers such that each is less than the sum of the other two. Prove that each of these numbers is greater than the distance between the other two, and that

$$a+b-c < \frac{2ab}{c}.$$

Deduce that if x, y, z are complex numbers which are not collinear then

$$|x-y| + |y-z| - |z-x| < 2\left|\frac{(x-y)(y-z)}{z-x}\right|.$$

116. For complex numbers u and v, prove that the following statements are equivalent to one another:

(i) u and v are not collinear with 0;

(ii) for each complex number z there are real numbers ξ and η, uniquely determined by z, such that $z = \xi u + \eta v$;

(iii) $\operatorname{im}(u/v)$ is positive or negative.

Suppose that u and v satisfy these conditions and that, moreover, $|u| = |v| = 1$ and $\operatorname{re}(u/v) = 0$. If $z = \xi u + \eta v$, express $|z|$ in terms of ξ and η; and if $z' = \xi' u + \eta' v$ and

$$zz' = \xi'' u + \eta'' v,$$

express ξ'' and η'' in terms of ξ, η, ξ', η' (distinguishing the cases in which $u/v = \pm i$).

117. Let C_1 and C_2 be circles in the complex plane, with centres c_1 and c_2, and radii r_1 and r_2, respectively. Show that unless $c_1 = c_2$ and $r_1 = r_2$, there are at most 2 points common to C_1 and C_2, and that the number of common points is 2 if and only if

$$|r_1 - r_2| < |c_1 - c_2| < r_1 + r_2.$$

Show that every circle in the complex plane has a unique centre and a unique radius.

118. Define the term *basis* for the system of 3-dimensional real vectors. Explain how the idea of a basis can be supplemented by the ideas of *right-handedness* and *left-handedness*, and how these ideas are related to the theory of determinants.

119. Prove *the basis theorem* for 3-dimensional real vectors, which states that a set is a basis if and only if it consists of 3 linearly independent vectors.

Discuss the possibility of generalizing this theorem to the case of *n*-dimensional vectors over an arbitrary field.

120. Discuss the relations between the concepts of *length*, *distance*, *scalar product*, and *orthogonality*, for 3-dimensional real vectors. In particular, show how we can calculate the scalar product of two vectors if we know the distance between them and either their lengths or the length of their sum, and show how orthogonality can be characterized in terms of 'the Pythagoras property'. Explain what is meant by an *orthonormal basis*.

Discuss the extent to which these ideas are available for *n*-dimensional real vectors, and for vectors over a field which may not be the real field.

121. Prove *Lagrange's identity*, which states that, for any vectors **p, q, r, s,**

$$(p \wedge q).(r \wedge s) = p.r\, q.s - q.r\, p.s.$$

Deduce that

$$(p \wedge q)^2 = p^2 q^2 - (p.q)^2,$$

and hence that

$$|p \wedge q| \leqslant |p|\,|q|,$$

equality holding if and only if **p** and **q** are othogonal to each other.

122. Let **u** and **v** be non-zero vectors, and let L and M be lines whose directions are specified by **u** and **v** respectively. Let **a** be any point of L, let **b** be any point of M, and suppose that L and M are not parallel. Show that the vectors **u, v, u ∧ v** constitute a basis, and that there is a unique line N intersecting both L and M orthogonally. Prove that the distance between the points of intersection of N with L and with M is

$$\frac{(a-b).u \wedge v}{|u \wedge v|},$$

and that this is less than the distance between **a** and **b** unless **a** and **b** are the points of intersection.

123. A *projective plane* is a set with a class of subsets called *lines* such that

(i) any two (distinct) lines have exactly one point in common;

(ii) any two (distinct) points belong to exactly one line;

(iii) there is a subset consisting of 4 points of which no 3 are collinear (belong to the same line).

Show that the complex plane, with lines defined in the usual way ('line' meaning 'straight line'), is not a projective plane.

For 3-dimensional real vectors $\mathbf{u}$, $\mathbf{v}$, both different from $\mathbf{0}$, let '$\mathbf{u} \sim \mathbf{v}$' mean that $\mathbf{u}$ is a scalar multiple of $\mathbf{v}$. Show that $\sim$, thus defined, is an equivalence relation, and that the corresponding equivalence classes constitute a projective plane if a 'line' is defined to be a set of equivalence classes whose union, together with $\mathbf{0}$, is a plane. The projective plane obtained in this way is called *the real projective plane*.

Show that the Cartesian plane can be embedded in the real projective plane by identifying each point (x, y) with the equivalence class that contains the vector $(x, y, 1)$, and that this embedding fills the real projective plane except for the points of a certain line, Λ_0 say. If we call Λ_0 *the line at infinity* corresponding to the given embedding, it is appropriate to call the intersection of Λ_0 with any other line Λ in the projective plane *the point at infinity on* Λ: show that the sets $\Lambda \setminus \Lambda_0$, obtained by deleting the point at infinity on each such line Λ, are identified by the embedding with the lines in the Cartesian plane, lines that have a common point at infinity corresponding in this way to parallel lines.

(The notion of parallelism, which is a main theme in 'affine geometry', is absent from 'projective geometry'—as is the notion of distance, which is essential to 'metrical geometry'; thus projective geometry differs fundamentally from Euclidean geometry, which has both affine and metrical aspects. The construction of the real projective plane in terms of 3-dimensional vectors—according to which a 'point' is a class of equivalent ordered triads of 'homogeneous coordinates'—can be used to construct a projective plane over any field; the complex projective plane is of particular interest, especially in the algebraic theory of 'conics'. Whatever the field, if the *direction of a line L* in the corresponding 3-dimensional space is defined to be the class of all lines parallel to L, the corresponding projective plane can be identified with the set of all such directions. If the field is totally ordered, the *direction of a non-zero vector* $\mathbf{u}$ is the set $\{\lambda \mathbf{u}: \lambda > 0\}$, and the *opposite* direction is the direction of $-\mathbf{u}$; so that the point represented by $\mathbf{u}$ in the corresponding projective plane is the union of these two directions. If the field is that of the real numbers, the directions of the non-zero vectors can be identified in an obvious way with the points of a sphere, opposite directions corresponding to antipodal points; thus the real projective plane can be constructed from a sphere by identifying antipodal points—a type of construction often used in topology. The various kinds of algebraic geometry that are largely concerned with lines can be regarded as geometrical aspects of linear algebra; they are known collectively as 'linear geometry'.)

124. Prove that a polynomial equation of degree n with coefficients in a given field cannot have more than n roots in the field.

Give an example of a commutative ring having distinct non-zero elements a and b such that $ab = 0$. Show that a quadratic equation with coefficients in such a ring can have more than 2 roots in the ring.

125. Let R be a ring such that if $x \in R$ and $x \neq 0$ there are elements y and z of R (possibly equal) such that $xy \neq 0$ and $zx \neq 0$. Let

$$RR = \{xy: x \in R, y \in R\},$$

and let f be a function which maps RR into R in such a way that if x, y, z are any elements of R then

$$xf(yz) = f(xyz) = f(xy)z.$$

Prove that, for any elements u, x, y, z of R,

$$f(uxy + uxz) = f(uxy) + f(uxz),$$

and hence that $\qquad f(xy + xz) = f(xy) + f(xz).$

Show that if we replace multiplication in R by the operation

$$(x, y) \rightarrow f(xy)$$

then we obtain a ring, R_f say. Show that if u is an element of R such that $ux = xu$ for every x in R then f could be, for example, the function $xy \rightarrow uxy$; and that if, further, R has a unit element and u has a reciprocal v in R then, with this choice of f, R_f has v as unit element.

Suppose that R consists of all real-valued functions on the open interval $]0, 1[$, and that $x \in R$ if and only if $\{x(t)/t: 0 < t < 1\}$ is bounded. Show that with addition and multiplication defined pointwise $((x+y)(t) = x(t) + y(t)$ and $(xy)(t) = x(t)y(t))$, R is a ring with no unit element, and that f can be chosen so that the ring R_f has a unit element.

126. Let f be a monotonic function which is additive in that

$$f(x+y) = f(x) + f(y)$$

for any real numbers x, y. Prove that $f(x) = xf(1)$ for every real number x.

127. Let C be a circle in the complex plane, with centre c and radius r. Prove that $z \in C$ if and only if

$$zz^* - c^*z - cz^* + cc^* = r^2.$$

Let f be the function defined as follows:

$$f(z) = \frac{1}{z} \quad (z \neq 0).$$

Prove that if $0 \notin C$ then $f(C)$ is a circle. If $0 \in C$, what kind of a set is $f(C \setminus \{0\})$?

128. Let f be a function which maps the complex plane one-to-one into itself in such a way that if L is any straight line then $f(L)$ is a straight line. Prove that if L and L' are parallel lines then $f(L)$ is parallel to $f(L')$.

Let $g(z) = f(z) - f(0)$, for every complex number z. By considering a parallelogram with a vertex at 0, prove that if z_1 and z_2 are complex numbers which are not collinear with 0 then

$$g(z_1 + z_2) = g(z_1) + g(z_2);$$

and, by considering a suitable configuration, deduce that this equation holds for any complex numbers z_1, z_2 (that is, that g is additive).

Suppose that f has the further property that it preserves betweenness, in the sense that if a, b, c are collinear points and b is between a and c then $f(b)$ is between $f(a)$ and $f(c)$. Prove that

$$g(\alpha z) = \alpha g(z)$$

for any real number α and any complex number z.

129. Let f be a function which maps the complex plane into itself in such a way that

$$f(z_1 + \alpha z_2) = f(z_1) + \alpha f(z_2)$$

for any real number α and any complex numbers z_1, z_2. Prove that $f(0) = 0$, and that there are real numbers a, b, c, d such that, for every complex number z,
$$\mathrm{re}\, f(z) = a\,\mathrm{re}\, z + b\,\mathrm{im}\, z,$$
$$\mathrm{im}\, f(z) = c\,\mathrm{re}\, z + d\,\mathrm{im}\, z.$$

Deduce that f is a one-to-one mapping if and only if it has the following property: for each complex number z there is a complex number w such that $f(w) = z$.

130. If f is a scalar-valued function of a vector variable, a *level surface* for f is a set $\{\mathbf{r}: f(\mathbf{r}) = c\}$, where c is a scalar which, with f, identifies the surface; and f is said to be *linear* if

$$f(\mathbf{r} + \alpha \mathbf{s}) = f(\mathbf{r}) + \alpha f(\mathbf{s})$$

for any scalar α and any vectors $\mathbf{r}$, $\mathbf{s}$. Prove that f is linear if and only if there is a vector $\mathbf{v}$ such that, for every vector $\mathbf{r}$,

$$f(\mathbf{r}) = \mathbf{v}.\mathbf{r}.$$

Deduce that if f is linear then its level surfaces are the planes parallel to some given plane, and that the converse proposition is false.

Prove that the level surfaces of f are the spheres with centre $\mathbf{0}$ if and only if there is a real-valued function g, of a non-negative real variable, such that, for every vector $\mathbf{r}$,

$$f(\mathbf{r}) = g(|\mathbf{r}|).$$

131. A vector-valued function f of a vector variable is said to be *linear* if

$$f(\mathbf{r} + \alpha\mathbf{s}) = f(\mathbf{r}) + \alpha f(\mathbf{s})$$

for any scalar α and any vectors $\mathbf{r}$, $\mathbf{s}$. Prove that f is linear if and only if there is a linear function g which is *adjoint* to f in the sense that, for any vectors $\mathbf{r}$, $\mathbf{s}$, $$f(\mathbf{r}).\mathbf{s} = \mathbf{r}.g(\mathbf{s});$$

show that such an adjoint function g is uniquely determined by f, and that f is adjoint to g.

Prove that if f is linear and E is a set of vectors such that $f(E)$ consists of linearly independent vectors, then E consists of linearly independent vectors. Prove also that the following statements about a linear function f are equivalent to one another:

(i) f is a one-to-one mapping;

(ii) if $f(\mathbf{r}) = \mathbf{0}$ then $\mathbf{r} = \mathbf{0}$;

(iii) if E is a set of linearly independent vectors then so is $f(E)$;

(iv) if E is a basis then so is $f(E)$;

(v)   for every vector $\mathbf{r}$ there is a vector $\mathbf{s}$ such that $f(\mathbf{s}) = \mathbf{r}$.

Show that these statements are true if f is a linear function such that $|f(\mathbf{r})| = |\mathbf{r}|$ for every vector $\mathbf{r}$, and show that this last property is equivalent, for a linear function f, to the property that if E is an orthonormal basis then so is $f(E)$.

Suppose that a basis $\mathbf{u}$, $\mathbf{v}$, $\mathbf{w}$ is given, and let $\mathbf{u}'$, $\mathbf{v}'$, $\mathbf{w}'$ be any vectors. Show that there is a unique linear function f such that $f(\mathbf{u}) = \mathbf{u}'$, $f(\mathbf{v}) = \mathbf{v}'$, $f(\mathbf{w}) = \mathbf{w}'$. When $\mathbf{u}'$, $\mathbf{v}'$, $\mathbf{w}'$ are expressed as linear combinations of $\mathbf{u}$, $\mathbf{v}$, $\mathbf{w}$, the coefficients form an array of 9 scalars (a 3-by-3 matrix) which characterizes this linear function in terms of the given basis: show that if the basis is orthonormal then the 9 scalars are the scalar products of $\mathbf{u}$, $\mathbf{v}$, $\mathbf{w}$ with $\mathbf{u}'$, $\mathbf{v}'$, $\mathbf{w}'$, and are the same as the corresponding scalars for the adjoint function, though permuted; and consider the corresponding determinants.

(The linear functions considered here and in **130** are examples of 'linear transformations', which are the chief objects of study in linear algebra. A linear transformation whose domain includes its range—as here—is called a *linear operator*, or *vector-space endomorphism*; one whose range consists of scalars—as in **130**—is called a *linear functional*. See **73, 75, 128, 129**.)

132. Among the real-valued functions that have a given set of real numbers as domain, we distinguish those that are monotonic, including those that are increasing, non-decreasing, and so on; these classifications apply in particular to infinite sequences of real numbers (which are real-valued functions having the set of all positive integers as domain). A *subsequence* of an infinite sequence $u_1, u_2, u_3, \ldots$ is an infinite sequence $v_1, v_2, v_3, \ldots$ such that $v_k = u_{n_k}$ for each positive integer k, where $n_1, n_2, n_3, \ldots$ is an increasing sequence of positive integers. Prove that if an infinite sequence of real numbers has no non-decreasing subsequence then it has a decreasing subsequence.

An infinite sequence $u_1, u_2, u_3, \ldots$ of real numbers or of complex numbers is said to be *convergent* if there is a number u such that, for any positive number ϵ, $|u_n - u| < \epsilon$ provided that n is greater than some integer which can be chosen when ϵ has been chosen. A simple argument of a type commonly used in analysis (involving an appeal to the triangle inequality) shows that there cannot be more than one such number u for a given sequence; if there is one it is called the *limit* of the sequence. Prove that all subsequences of a convergent sequence are convergent, with the same limit. Prove also that a sequence of real numbers is *bounded* (has a bounded range) if it is convergent; prove that the converse of this proposition is false, but that every bounded monotonic sequence is convergent, its limit being its supremum or its infimum.

A *Cauchy sequence*, of real numbers or of complex numbers, is an infinite sequence $u_1, u_2, u_3, \ldots$ such that, for any positive number ϵ, $|u_m - u_n| < \epsilon$ provided that m and n are greater than some integer which can be chosen when ϵ has been chosen. Prove that every convergent sequence is a Cauchy sequence. Prove also that every Cauchy sequence of real numbers is convergent, its limit being that of a bounded monotonic subsequence.

The assertion that every Cauchy sequence is convergent is known as *the general principle of convergence*. Having established this principle for sequences of real numbers, deduce that it is valid also for sequences of complex numbers, and for sequences of 3-dimensional real vectors when convergence of such sequences is appropriately defined.

Show that for any real number x an infinite sequence $\{u_n\}$ of rational numbers can be defined so that, for each n,

$$|u_{n+2}-u_{n+1}| = \tfrac{1}{2}|u_{n+1}-u_n| \quad \text{and} \quad |u_n-x| \leqslant |u_{n+1}-u_n|$$

(as in an ancient—and modern—method of finding approximate square roots). Show that $\{u_n\}$ is then convergent, with limit x. Hence show that the general principle of convergence is not valid within the system of rational numbers. (The principle is in fact valid for a totally ordered field if and only if the field is complete. Since it is valid for some systems that are not totally ordered fields, it can be used to generalize the notion of completeness. The system of real numbers can be defined in terms of Cauchy sequences of rational numbers, as was first shown, late in the nineteenth century, by the German mathematician Georg Cantor. A. L. Cauchy, French, worked much earlier in that century.)

133. Let f be a real-valued function which is non-increasing on a closed interval $[a, a+n]$, where n is a positive integer. Prove that

$$f(a+1)+f(a+2)+\ldots+f(a+n)$$
$$\leqslant \int_a^{a+n} f(x)\,dx \leqslant f(a)+f(a+1)+\ldots+f(a+n-1).$$

Prove that, for any integer n greater than 2,

$$\frac{1}{2\log 2}+\frac{1}{3\log 3}+\ldots+\frac{1}{n\log n} \geqslant \log\log n - \log\log 2.$$

134. Let f and g be real-valued functions defined on an interval I, let c be a point of I, and suppose that there is a real number λ such that

$$\lim_{x\to c} f(x) = \lambda = \lim_{x\to c} g(x).$$

Prove or disprove the assertion that $f(c)$ and $g(c)$ are necessarily equal. Prove that if h is a function such that

$$f(x) \leqslant h(x) \leqslant g(x)$$

for every point x of I, then $\lim_{x\to c} h(x) = \lambda$.

135. A function f satisfies the following conditions:

$$f'(x)+2xf(x) = 0 \quad (x > 0), \qquad 0 \leqslant f(1) \leqslant 1.$$

Prove that if $x \geqslant 1$ then

$$0 \leqslant \int_1^x f(t)\,\mathrm{d}t < \tfrac{1}{2}.$$

136. Investigate the local maxima and the local minima of the function f defined as follows:

$$f(x) = \begin{cases} 0 & \text{if } x < 0, \\ e^{-x}\sin x & \text{if } x \geqslant 0. \end{cases}$$

137. A certain real-valued function g is known to be such that if we put $f = g$ we have a solution of the differential equation

$$f''(t) + f(t) = 0,$$

where t is any real number. Let $h = g'$. Show that $h' = -g$, that the function $g^2 + h^2$ has a constant value, and that we have a solution of the differential equation if we put

$$f = \alpha g + \beta h,$$

where α and β are any constants (real or complex).

Now suppose that f is any solution of the differential equation. Show that each of the functions $fg + f'h$ and $fh + f'g$ has a constant value. Assuming that g does not have a constant value, deduce the existence of constants α and β such that $f = \alpha g + \beta h$.

Show that for any real number u there are constants α and β such that

$$g(t+u) = \alpha g(t) + \beta h(t)$$

for every real number t. Evaluate α and β in terms of the numbers $g(0)$, $h(0)$, $g(u)$, $h(u)$, assuming that g does not have a constant value.

138. Let k be a non-zero real number. Prove that a function f satisfies the differential equation

$$f''(t) - k^2 f(t) = 0, \tag{1}$$

for every real number t, if and only if $f(t) = g(kt)$ for every t, where g is a function that satisfies the differential equation

$$g''(t) - g(t) = 0. \tag{2}$$

Prove also that g satisfies equation (2) if and only if it satisfies a differential equation

$$g'(t) + g(t) = h(t) \tag{3}$$

in which h is a function that satisfies the differential equation

$$h'(t) - h(t) = 0. \tag{4}$$

Show that if h is a non-vanishing solution of equation (4) then equation (3) is equivalent to

$$(hg)'(t) = h^2(t).$$

Hence obtain the general solution of equation (1) in terms of the exponential function and two arbitrary constants.

139. Consider the differential equation

$$f''(t) + 2bf'(t) + cf(t) = 0,$$

where b and c are real constants and t is any point of a given interval I which has more than one point. Show that f satisfies this equation if and only if

$$f(t) = e^{-bt}g(t) \quad (t \in I),$$

where g is a function such that

$$g''(t) + (c - b^2)g(t) = 0 \quad (t \in I).$$

Show that if $c > b^2$ then g satisfies this last equation if and only if

$$g(t) = h(\omega t) \quad (t \in I),$$

where $\omega = (c - b^2)^{\frac{1}{2}}$ and h is a function such that

$$h''(t) + h(t) = 0 \quad (t \in I).$$

Hence obtain the general solution of the given differential equation for the case in which $c > b^2$. Show that if $c = b^2$ the general solution is

$$f(t) = (\alpha t + \beta)e^{-bt} \quad (t \in I),$$

where α and β are constants. Discuss the case in which $c < b^2$.

140. Consider the differential equation

$$f''(t) + 2bf'(t) + cf(t) = g(t),$$

where b and c are constants, t is any point of a given interval I which has more than one point, and g is a given function defined on I. Let f_1 be a particular solution of this equation. Show that f is a solution if and only if

$$f(t) = f_1(t) + f_0(t) \quad (t \in I),$$

where f_0 is a solution for the case in which

$$g(t) = 0 \quad (t \in I).$$

Show that if g is a polynomial function then there is a polynomial function which is a particular solution of the differential equation. Discuss the case in which

$$g(t) = e^{kt}p(t) \quad (t \in I),$$

where k is a real constant and p is a polynomial function.

The differential equation considered here is typical of what are called *linear differential equations of the second order with constant coefficients.* (Such equations arise in the study of simple dynamical systems having inertia, resilience, and damping.) Consider reasons for calling them 'linear'.

FOURTH COLLECTION OF
EXERCISES (141–200)

CONSISTING OF SIX ADVANCED-LEVEL
EXAMINATION PAPERS

(Candidates were allowed 3 hours for each paper of 10 questions,
and in that time they were expected to answer 7 questions.)

First paper 1966

141. Explain how the empty set $\varnothing$ plays the role of a zero in the algebra
of sets.

A, B, and C are subsets of some given set. Prove that if $C \cap B = \varnothing$ then

$$(A \cup C) \cap (A \cup B) = A.$$

Determine whether or not the converse statement is true.

142. (i) If the decimal representation of the natural number n is
$d_r d_{r-1} \dots d_0$, then 11 divides n if and only if 11 divides

$$d_0 - d_1 + \dots + (-1)^r d_r.$$

State and prove the corresponding proposition in any scale of notation.

(ii) The binary representation of n_r consists of r ones followed by
$r + 1$ zeros, and ends with a single one. Prove that n_r is a perfect square,
and say how its positive square root would be represented in binary
notation.

143. $P(x)$ and $Q(x)$ are polynomials in an indeterminate x, with coefficients in a given field. Show that there is a polynomial $R(x)$ which divides
$P(x)$ and $Q(x)$ and is divisible by every common factor of $P(x)$ and $Q(x)$,
and that there exist polynomials $A(x)$ and $B(x)$ such that

$$R(x) = A(x) P(x) + B(x) Q(x),$$

the coefficients of $A(x)$, $B(x)$, and $R(x)$ being in the same field as those of
$P(x)$ and $Q(x)$. Find polynomials $R(x)$, $A(x)$, and $B(x)$, which satisfy the
conditions above, if $P(x) = x^3 + 1$, $Q(x) = x^5 + 1$.

144. (i) Sketch the graphs of

(a) $e^x - x$,

(b) $\tan^{-1}(x^{-1})$ for $x > 0$.

(*Note:* Accurate graphs on squared paper are *not* required, but a sketch must indicate the important features of the curve being considered.)

(ii) For what real values of x has the expression

$$\frac{1}{1 + 1/(2+x)}$$

no meaning? For real values of x for which the expression has meaning, state, with reasons but without using differentiation, whether it represents an increasing or a decreasing function of x.

145. The line l_1 passes through the two points $(-1, 3, -5)$ and $(2, 7, 0)$. The line l_2 lies in the two planes $x - 2y + z = 0$ and $y = 3$. Represent each line in the form $\mathbf{r} = \mathbf{a} + t\mathbf{b}$, where $\mathbf{r} = (x, y, z)$, $\mathbf{a}$ and $\mathbf{b}$ are numerical vectors, and t is a scalar.

Obtain a similar representation for the line that cuts l_1 and l_2 orthogonally.

146. (i) Find the real and imaginary parts of $(3i - 5)/(4 + i)$.

(ii) Find the real and imaginary parts of the roots of the equation

$$z^2 - (4 + 2i)z + 6 = 0.$$

(iii) Find that fifth root of $-4 - 4i$ which has the greatest real part.

147. (i) Differentiate $(\tan x)/(\sin x + \cos x)$ with respect to x.

(ii) Show that the nth derivative of $e^{ax} \sin bx$ can be written in the form $(a^2 + b^2)^{\frac{1}{2}n} e^{ax} \sin(bx + n\alpha)$,

and explain how the number α can be found if a and b are given.

148. Evaluate the integrals

$$\int_0^1 \frac{x^3 + 3x^2 + x + 1}{(x+1)^2(x^2+1)}\,dx, \quad \int_0^2 \frac{dx}{\sqrt{\{(3-x)(1+x)\}}}, \quad \int_0^1 x\,e^x\,dx.$$

By considering bounds for the second integrand show that $3 < \pi < 2\sqrt{3}$.

149. If, for all real values of x,

$$(x - 1)y''(x) - xy'(x) + y(x) = 0,$$

show that if $u(x) = e^{-x}y(x)$ and $v(x) = u'(x)$ then $v(x)$ satisfies a first-order differential equation. Solve this differential equation, and hence obtain expressions for $u(x)$ and $y(x)$ involving x and arbitrary constants.

150. Starting from the equations

$$\frac{d}{d\theta}\sin\theta = \cos\theta, \quad \frac{d}{d\theta}\cos\theta = -\sin\theta, \quad \sin 0 = 0, \quad \cos 0 = 1,$$

prove the addition formula giving $\sin(\theta + \varphi)$ in terms of sines and cosines of θ and φ. Express $\sin 3\theta$ in terms of $\sin\theta$, and hence find the value of $\sin\frac{1}{3}\pi$.

Second paper 1966

151. Use Euclid's algorithm to find integers a and b such that

$$17a + 29b = 1.$$

Hence find a general expression for the integers x such that the two congruences $x \equiv 2 \pmod{17}$ and $x \equiv 5 \pmod{29}$ hold simultaneously.

152. Each of the statements (i) to (iv) below defines a relation between two non-zero integers a and b. For each, determine whether or not the relation is an equivalence relation, and if it is, describe the equivalence classes that are defined by it.

  (i) $a \sim b$ if and only if $|a - b| \leqslant 1$.

  (ii) $a \sim b$ if and only if $(a - b)$ is divisible by 7.

  (iii) $a \sim b$ if and only if $a \geqslant b$.

  (iv) $a \sim b$ if and only if there exists an integer n divisible by both a and b.

153. Define the terms *supremum* and *infimum* with reference to a bounded set of real numbers. Why are these terms not applicable with reference to a set of complex numbers?

State the supremum and infimum of the set

$$\{n^3 4^{-n} : n = 1, 2, 3, \ldots\},$$

and prove your statement.

154. p, q, and r are three 3-dimensional real vectors such that

$$\mathbf{p} \cdot \mathbf{p} \neq 0, \quad \mathbf{q} \cdot \mathbf{q} \neq 0, \quad \mathbf{r} \cdot \mathbf{r} \neq 0, \quad \mathbf{q} \cdot \mathbf{r} = 0, \quad \mathbf{r} \cdot \mathbf{p} = 0, \quad \mathbf{p} \cdot \mathbf{q} = 0.$$

Prove that p, q, and r are linearly independent. State, with reasons, whether or not this is true if the given conditions hold except that (a) $\mathbf{p} \cdot \mathbf{p} = 0$, or (b) $\mathbf{p} \cdot \mathbf{q} \neq 0$.

155. (i) Explain what is meant by the *domain* and the *range* of a function.

Find the largest real domain for a real-valued function f represented by the equation

$$f(x) = \frac{\sqrt{x}}{1+\sqrt{x}}.$$

State, with reasons, what the range of f is with this domain.

(ii) Give necessary and sufficient conditions on the domains and ranges of the functions g and h in order that the composite function $g \circ h$ may be defined. Give examples of functions g and h (neither of which is the identity function $g(x) = x$) such that

(a) $g \circ h = h \circ g$,

(b) $g \circ h \neq h \circ g$.

156. Define the function log, and, from the definition, prove that $\log x^r = r \log x$ if r is a rational number.

Discuss whether or not it follows from this that

$$\log x^a = a \log x$$

for all real values of a.

157. By putting $\left(1+\frac{r}{n}\right)^{-1}\left(1+\frac{r-1}{n}\right)^{-1}$

into partial fractions prove that, for any positive integer n,

$$\frac{1}{1(1+1/n)} + \frac{1}{(1+1/n)(1+2/n)} + \cdots + \frac{1}{(1+(n-1)/n)\,2} = \frac{n}{2}.$$

By dissecting the interval $1 \leqslant x \leqslant 2$ into subintervals of equal length, and finding bounds for upper and lower Riemann sums, prove that

$$\int_1^2 x^{-2}\,dx = \tfrac{1}{2}.$$

158. Define the derivative $f'(x)$ of the function f at the point x.

Prove that if f is continuous on an interval containing c and x then

$$\frac{d}{dx}\int_c^x f(t)\,dt = f(x).$$

Differentiate $\displaystyle\int_x^{x^3} \frac{dt}{\sqrt{(1+t^2)}}$ with respect to x.

159. Let f and g be functions which are differentiable everywhere, and suppose that a and b are real numbers such that $a < b$, $g(a) \neq g(b)$. By consideration of the determinant

$$\begin{vmatrix} f(x) & g(x) & 1 \\ f(a) & g(a) & 1 \\ f(b) & g(b) & 1 \end{vmatrix}$$

show that there exists a real number c such that $a < c < b$ and

$$\frac{f(a)-f(b)}{g(a)-g(b)} = \frac{f'(c)}{g'(c)}.$$

Explain the flaw in the argument 'By the mean-value theorem,

$$f(a)-f(b) = (a-b)f'(c) \quad \text{and} \quad g(a)-g(b) = (a-b)g'(c),$$

whence the result follows.'

Evaluate the limit of $(e^x - 1)/x$ as x tends to 0.

160. Let U be a function defined on the closed interval $-1 \leqslant x \leqslant 1$ by the equation

$$U(x) = \int_0^x \frac{2\,dt}{1+t^2}.$$

Show that U has an inverse. Define, for all values of x, $\sin x$ in terms of this inverse function, and prove that $\sin$ is a continuous function.

Special paper 1966

161. Define the terms *Cartesian product* and *one-to-one correspondence*.

What can be deduced about the number of elements in the finite sets A and B if $A \times B$ is in one-to-one correspondence (a) with A, or (b) with $A \cap B$?

162. Explain how the ring of integers is constructed by means of classes of equivalent ordered pairs of natural numbers. (The principal steps in the construction must be stated, but the results of simple algebraic steps may be quoted without giving the working in full.)

163. Prove that $\sqrt{3} - \sqrt{2}$ is not a rational number.

Let R be the ring of polynomials with rational coefficients, in a single indeterminate x, and let F be the set of numbers obtained from R by giving x the value $\sqrt{2}$. Prove that F is a field which is not the whole field of real numbers and which contains the field of rational numbers as a proper subfield.

164. Let C be the set of complex numbers z such that

$$|z - 12| = 2|z - 9i|.$$

Find a complex number w_0 and a positive real number μ such that if $z = w - w_0$ then C is the set of complex numbers w such that $|w| = \mu$.

Give a geometrical interpretation of this result by identifying the field of complex numbers with the Cartesian plane.

165. Define the *vector product* of two 3-dimensional vectors. Prove that $a \wedge (b \wedge c)$ is a linear combination of b and c.

Is vector multiplication (i) commutative, (ii) associative, (iii) distributive with respect to addition? Give proofs.

166. Three polynomials P, Q, and R, of degrees $2n - 1$, $2n$, and $2n + 1$, respectively, are such that

(i) $R(x) = 3xQ(x) - P(x)$ for all real x,

(ii) $P(1) = Q(1) = 1$,

(iii) $P(x)$ contains only odd powers of x and $Q(x)$ contains only even powers of x,

(iv) the equation $Q(x) = 0$ has $2n$ real roots in the interval $]-1, 1[$, and between each two consecutive roots lies a root of the equation $P(x) = 0$.

Prove that the equation $R(x) = 0$ has $2n + 1$ real roots in the interval $]-1, 1[$, and that between each two consecutive roots lies a root of the equation $Q(x) = 0$.

167. Let the function t be defined, for real values of x such that $-1 \leqslant x \leqslant 1$, by the conditions

$$t(0) = 1,$$
$$t(x) = \frac{1}{x[1/x]} \quad \text{if } x \neq 0.$$

($[x]$ denotes the greatest integer not greater than x.) Prove that t is continuous at $x = 0$, and determine whether or not it is differentiable there.

168. For any real number x, let $f(x)$ denote x minus the largest integer not greater than x, and let $g(x) = f(\sqrt{x})$. State whether or not

(a) g is continuous on the interval $1 \leqslant x \leqslant 3$,

(b) g is continuous on the interval $1 \leqslant x \leqslant 4$,

(c) $\displaystyle\int_{1}^{4} g(x)\,dx$ exists as a Riemann integral and equals $\frac{5}{3}$.

Give proofs.

169. Show that if f is a differentiable function such that $f'(x) = f(x)$ for all real x, then $e^{-x}f(x)$ is constant.

If $g(0) = 0$ and $g'(x) \geq g(x)$ for $x \geq 0$, prove that $g(x) \geq 0$ for $x \geq 0$.

If h is a function such that $h(x) \leq 1$ for $0 \leq x \leq 1$, and if y satisfies the differential equation

$$y'(x) = y(x) + h(x),$$

with $y(0) = 1$, prove that $y(1) \leq 2e - 1$.

170. Give a definition of the function e^x, and prove that

$$\frac{\mathrm{d}}{\mathrm{d}x}e^x = e^x \quad \text{and} \quad e^0 = 1.$$

Evaluate $\displaystyle\int_0^1 x^n e^{-x}\,\mathrm{d}x$ and hence or otherwise calculate the numerical value of e^1 with an error not more than 10^{-2}.

First paper 1967

171. If A and B are subsets of a set X, prove that

$$X \setminus (A \cup B) = (X \setminus A) \cap (X \setminus B).$$

State a similar expansion of $X \setminus (A \cap B)$.

If X is the set of positive integers, A is the set of positive even integers, B is the set of squares of integers, and C is the set of positive multiples of 3, give five elements of $\quad X \setminus \{A \cup (B \cap C)\}.$

172. If M is an odd prime number show that the numbers 0^2, 1^2, ..., $(M-1)^2$ give exactly $(M+1)/2$ different remainders on division by M. If M is odd but not a prime, state, with reasons, whether or not the number of different remainders is necessarily $(M+1)/2$, and, if not, whether it may be larger.

Find the integers between 1 and 12 which are not congruent to squares modulo 13.

173. Show that in the field of rational numbers the equations

$$
\begin{aligned}
4x + 3y + z &= 1, \\
2x + y + 4z &= 1, \\
x \qquad - 5z &= 1,
\end{aligned}
$$

have only one solution, but that in the field of integers modulo 7 they have more than one solution. Find all the solutions in the field of integers modulo 7.

Give an example of a pair of linear equations which has no solution in the field of rational numbers but which has solutions in the field of integers modulo 3.

174. (a) Prove algebraically that, if z_1 and z_2 are any two complex numbers, $|z_1 + z_2| \leqslant |z_1| + |z_2|$. Give the geometrical interpretation of this when the complex number $x + iy$ is identified with the point (x, y) of the Cartesian plane.

(b) Find the real and imaginary parts of $(7 - i)/(2 - i)$, and of the numbers z such that $z^3 = -8i$.

175. Let $\mathbf{r}$ denote the vector (x, y, z), and let $\mathbf{a}$ be a given non-zero vector and b a scalar. Explain the geometrical significance of the equation $\mathbf{r} \cdot \mathbf{a} = b$.

Let π_1 be the plane through $(1, 2, 3)$ with normal in the direction $(1, 1, 1)$, and let π_2 be the plane through the points $(0, 0, 0)$, $(1, 0, 0)$, and $(0, 1, 1)$. Find the direction of the line of intersection of π_1 and π_2; find also a point on this line of intersection.

176. Sketch the graphs of the functions

(a) $2x/(1 + x^2)$,

(b) $(x + 2)/(x^2 - 1)$ $(x \neq 1, -1)$,

showing, in particular, where maximum and minimum values occur.

*Accurate graphs on squared paper are **not** required.*

177. Define $\log x$ for $x > 0$, and prove that

$$\log xy = \log x + \log y.$$

Prove that, for $1 \leqslant x \leqslant 1 + 2A$,

$$\frac{1}{1 + A} - \frac{(x - A - 1)}{(1 + A)^2} \leqslant \frac{1}{x} \leqslant 1 - \frac{x - 1}{1 + 2A},$$

and hence prove that

$$\frac{2A}{1 + A} \leqslant \log(1 + 2A) \leqslant \frac{2A(1 + A)}{1 + 2A}.$$

Determine whether or not these inequalities for $\log(1 + 2A)$ hold for $-\frac{1}{2} < A < 0$.

178. Explain concisely the principle of mathematical induction. Use induction to prove that

$$\frac{d^n}{dx^n}\left(\frac{1}{1+x^2}\right) = (-1)^n n!(1+x^2)^{-\frac{1}{2}(n+1)}\cos\{(n+1)\tan^{-1}x - \tfrac{1}{2}n\pi\}.$$

179. (a) Evaluate $\displaystyle\int_1^2 \frac{2x^3+x+1}{x^2(x^2+1)}\,dx.$

(b) Show that $\displaystyle\int_{-\sqrt{\frac{2}{3}}}^{\sqrt{\frac{2}{3}}} \frac{dx}{(\sqrt{3}-x\sqrt{2})\sqrt{(1-x^2)}} = \tfrac{1}{2}\pi.$

180. (a) If $dy/dx = y/x$ for all non-zero x, and $y = 1$ when $x = 1$, find y in terms of x for $x > 0$. Determine whether or not the information above is sufficient to determine the value of y when $x = -1$.

(b) If $dy/dx = 1 - \{2xy/(x^2+1)\}$ for all x, and $y = 1$ when $x = 1$, find y in terms of x for all real x.

Second paper 1967

181. Find, with proofs, the suprema and infima of the sets

$$\left\{\tfrac{1}{2}+(-1)^n\frac{n}{n^2+4}: n = 1,2,3,...\right\} \quad \text{and} \quad \{n.2^{-n}: n = 1,2,3,...\}.$$

It may be assumed that $r^n \to 0$ as $n \to \infty$ if $|r| < 1$.

182. Let $\mathbf{a}$, $\mathbf{b}$, $\mathbf{c}$ denote respectively the vectors $(1,2,3)$, $(1,0,-1)$, $(7,4,1)$. Show that $\mathbf{a}$, $\mathbf{b}$, $\mathbf{c}$ are linearly dependent.

Find all vectors $\mathbf{d}$ which are such that $\mathbf{a}$, $\mathbf{b}$, $\mathbf{d}$ are linearly dependent and $\mathbf{c}.\mathbf{d} = 0$ and $\mathbf{d}.\mathbf{d} = 1$.

183. Show that for two polynomials P and Q to have no common factor it is necessary and sufficient that there exist two polynomials A and B such that

$$\frac{1}{PQ} = \frac{A}{P} + \frac{B}{Q}.$$

(Any theorem used in the argument must be clearly stated, but need not be proved.)

Expand $\dfrac{1}{(x^2-1)(x^3-1)}$ in partial fractions.

184. Give the expression for ^nC_r in terms of factorials, where ^nC_r denotes the coefficient of $a^{n-r}b^r$ in the expansion of $(a+b)^n$ by the binomial theorem.

Prove that if p is a prime then p divides ^pC_r for $1 \leqslant r \leqslant p-1$. Hence prove that the divisibility of $2^m - 2$ by m is a necessary condition for m to be prime, and by consideration of $m = 341 = (2^{10} - 1)/3$ show that the condition is not sufficient.

185. Prove that if $a < b < c$ and the function f is continuous on the intervals $[a, b]$ and $[b, c]$ then it is continuous on $[a, c]$.

Prove that the function $|x|$ is continuous on any interval. Hence prove that if f and g are continuous on an interval I then so are $\max(f, g)$ and $\min(f, g)$, where $\{\max(f, g)\}(x)$ is the greater of $f(x)$ and $g(x)$ and $\{\min(f, g)\}(x)$ is the lesser of $f(x)$ and $g(x)$. (Standard theorems on the continuity of sums and differences may be quoted without proof.)

186. Find a, b, c, d so that the function f, defined by the conditions

$$f(x) = -x^2 \quad \text{for} \quad x < 0,$$
$$f(x) = a + bx + cx^2 + dx^3 \quad \text{for} \quad 0 \leqslant x \leqslant 1,$$
$$f(x) = x \quad \text{for} \quad x > 1,$$

has a continuous derivative for all x.

Determine whether or not f has an inverse.

187. Give, with proofs, the values of

$$\lim_{x \to 1} \frac{\log x}{x-1} \quad \text{and} \quad \lim_{x \to 0} \frac{1}{x[1/x]},$$

where $[y]$ means the greatest integer not greater than y.

188. Explain, in terms of Riemann sums, what is meant by the existence of $\int_a^b f(x)\,dx$. From this definition, prove that

$$\int_{a-k}^{b-k} f(x+k)\,dx = \int_a^b f(x)\,dx.$$

State, and prove similarly, a similar result for

$$\int_{a/m}^{b/m} f(mx)\,dx \quad (m > 0).$$

189. Prove that every solution of the differential equation $f''(\theta) + f(\theta) = 0$ is of the form $f(\theta) = A\sin\theta + B\cos\theta$, where A and B are constants.

Hence obtain the addition formulae for the functions sin, cos, and tan. Solve the equation $2\tan^{-1}(1/3) + \tan^{-1}(1/x) = \frac{1}{4}\pi$.

190. Explain what is meant by the statement 'the function

$$t - \sin t + i(1 - \cos t) \quad \text{for} \quad 0 < t < \pi$$

defines a simple differentiable arc'. Find the length of the arc so defined.

Special paper 1967

191. Let N be the set of natural numbers. Prove that the Cartesian product $N \times N$ can be put in one-to-one correspondence with N. Prove also that the positive rational numbers can be put in one-to-one correspondence with the natural numbers.

192. Write down the natural numbers less than 8 which have no common factor (other than 1) with 8, and show that they form an Abelian group under the operation of multiplication modulo 8.

Do the same thing with 8 replaced by 5; prove that the two groups that have been obtained are not isomorphic.

193. Find the number m which has a four-digit representation ending with 11 in both the decimal scale and the scale with base 7. Find also two other scales in which the representation of m ends with 11.

194. By using the identity $(a+b)^3 = a^3 + b^3 + 3ab(a+b)$, or otherwise, prove that there are no integers p, q, r, not all zero, such that

$$p \cdot 2^{\frac{2}{3}} + q \cdot 2^{\frac{1}{3}} + r = 0.$$

195. For ordered pairs of rational numbers, let addition and multiplication be defined by

$$(a, b) + (c, d) = (a+c, b+d),$$
$$(a, b) \cdot (c, d) = (ac+bd, ad+bc).$$

Prove that addition and multiplication are commutative and associative, and that multiplication is distributive with respect to addition.

Determine whether or not the pairs form a field with respect to these operations.

196. If z_1 and z_2 are two complex numbers, and $\overline{z_1}$ is the conjugate of z_1, prove that

$$|\overline{z_1}| = |z_1|, \quad |z_1 z_2| = |z_1| \, |z_2|.$$

Prove that the product of two numbers, each of which is the sum of squares of two unequal natural numbers, can be expressed as the sum of squares of two unequal natural numbers in at least two distinct ways ($a^2 + b^2$ and $b^2 + a^2$ counting as the same). Express 14,645 as the sum of squares of two natural numbers in four ways.

197. If **a**, **b**, **c** are any three vectors, prove that

$$\mathbf{a} \wedge (\mathbf{b} \wedge \mathbf{c}) = (\mathbf{a} . \mathbf{c})\mathbf{b} - (\mathbf{a} . \mathbf{b})\mathbf{c}.$$

Show that the condition $\mathbf{a} \wedge \mathbf{c} = 0$ is sufficient for

$$\mathbf{a} \wedge (\mathbf{b} \wedge \mathbf{c}) = (\mathbf{a} \wedge \mathbf{b}) \wedge \mathbf{c}.$$

Determine whether or not it is a necessary condition.

198. Let rational numbers be written in the form p/q, where p and q are integers with no common factor. Prove that in any interval there are rational numbers with q even and also rational numbers with q odd.

Let a function f be defined as follows:

$$f(p/q) = 1 \quad \text{if } q \text{ is even,}$$
$$f(p/q) = 0 \quad \text{if } q \text{ is odd,}$$
$$f(x) = 0 \quad \text{if } x \text{ is irrational.}$$

Prove that $\int_0^1 f(x)\,dx$ does not exist as a Riemann integral.

199. Defining e^x as the function inverse to $\log x$, prove that if f is a function such that $e^x > f(x)$ for $x > 0$ then

$$e^x > 1 + \int_0^x f(t)\,dt \quad \text{for} \quad x > 0.$$

Hence prove that

$$e^x > 1 + x + \frac{x^2}{2} + \ldots + \frac{x^n}{n!} \quad \text{for} \quad x > 0.$$

Prove that, corresponding to a given natural number n, there is a number $A(n)$ such that

$$e^x > x^n \quad \text{for} \quad x > A(n).$$

200. Prove that if f is continuous for $a \leqslant x \leqslant b$ and differentiable for $a < x < b$, then there is a number c such that $f(b) - f(a) = (b-a)f'(c)$ and $a < c < b$. (If Rolle's theorem is quoted then it must be proved.)

Prove that if f is a cubic function then c may be chosen so that

$$\frac{2a+b}{3} \leqslant c < b.$$

SOLUTIONS AND COMMENTS
FOR EXERCISES 1–50

1. The information conveyed is as follows:
   (i) B is contained in (is a subset of) A. Equivalently: A includes B; or, there is no element of B which is not also an element of A. (It is possible to say this in other ways, but one must be careful not to exclude the possibility that B may be empty. In symbols, we usually write $B \subseteq A$ or $A \supseteq B$—reserving $\subset$ and $\supset$ for occasions when we wish to assert that the inclusion is *strict*, giving the additional information that $A \neq B$.)
   (ii) The same as (i).
   (iii) A and B do not intersect. Equivalently: there is no element which belongs both to A and to B.
   (iv) A is the empty set and B is the empty set. Equivalently: A and B are both empty; or, neither A nor B is a proper set.
   (v) The same as (iii).
   (vi) A is contained in B.

   The explanations are simple. For instance, (iv) states that the elements, if any, that belong to A but not to B are precisely the elements that belong to B: since an element cannot both belong and not belong to B, we conclude that $B = \varnothing$; thus $A \setminus \varnothing = \varnothing$, and therefore $A = \varnothing$ (since $A \setminus \varnothing = A$ for any set A).

2. By definition, $A \cup B$ consists of those elements x for which
$$x \in A \quad \text{or} \quad x \in B$$
(the word 'or' being used here in the usual 'wide' sense, not the 'exclusive' sense). Thus, by the symmetry of the definition,
$$A \cup B = B \cup A.$$
To prove that $A \cup (B \cup C) = (A \cup B) \cup C$, suppose first that $x \in A \cup (B \cup C)$, so that $x \in A$ or $x \in B \cup C$. If $x \in A$ then $x \in A \cup B$ and hence $x \in (A \cup B) \cup C$; and if $x \in B \cup C$ then $x \in A \cup B$ or $x \in C$, so that again $x \in (A \cup B) \cup C$. Thus
$$A \cup (B \cup C) \subseteq (A \cup B) \cup C.$$

We can similarly demonstrate the opposite inclusion, and so establish the required equation.

If a is the single element of A then either (i) a of $A \cup B$ corresponds to a of $A \cup C$, or (ii) a of $A \cup B$ corresponds to, say, c of $A \cup C$ while a of $A \cup C$ corresponds to, say, b of $A \cup B$, where $c \in C$ and $b \in B$. In case (i) the remaining correspondences between the elements of $A \cup B$ and those of $A \cup C$ give a one-to-one correspondence between B and C, and in case (ii) we can let b correspond to c and then have the remaining correspondences as in case (i). (The hypothesis that A be disjoint from $B \cup C$ is not superfluous, as can be seen by letting $A = \{1\}$, $B = \{2\}$, and $C = A \cup B$. We can allow A to have more than one element, but not infinitely many: for example, if A consists of all the even numbers, B of one odd number, and C of two odd numbers, then $A \cup B$ and $A \cup C$ are in one-to-one correspondence but B and C are not.)

If A, B, and C are disjoint finite sets, having m, n, and p elements respectively, the above results show respectively that

$$m + n = n + m,$$

$$m + (n + p) = (m + n) + p,$$

and ($m = 1$ in the third case)

$$n = p \quad \text{if} \quad n + 1 = p + 1.$$

These are the fundamental laws of addition for natural numbers.

(The first two of these laws of addition are respectively the commutative law and the associative law; from the third we can deduce, by mathematical induction, the cancellation law, which states that $n = p$ if there is a natural number m such that $n + m = p + m$. The practice of using the symbol $+$ to denote addition of numbers seems to have originated, together with the use of $-$ for subtraction, towards the end of the fifteenth century, in Germany; it was adopted by the sixteenth-century Welsh mathematician Robert Recorde, who was the first person to write a noteworthy mathematical book in English, and the first to use $=$ as a mathematical symbol. The symbol $+$ is now used for various binary operations, but its use is generally confined to operations that are commutative and associative. The symbol $\cup$, denoting union of sets, was introduced, together with $\cap$ for intersection, by the nineteenth-century Italian mathematician G. Peano; but for much of the first half of the twentieth century $+$ was commonly used instead of $\cup$, the term 'sum', or 'logical sum', being used instead of 'union'. Peano's notation has now been universally adopted, and the expressions $A \cup B$ and $A + B$ can occur in the same context but with different meanings: if A and B are subsets of a set with a binary operation $+$ then, by definition,

$$A+B = \{a+b:\ a \in A,\ b \in B\};$$

and $A \setminus B$ is not written $A-B$. See **8, 15, 51, 52, 56, 57, 81, 84**.)

3. *The principle of mathematical induction* can be stated as follows. Suppose that for each natural number n $(1, 2, 3, ...)$ we have a certain proposition P_n: then P_n is true for every n if

(i) P_1 is true

and (ii) for every n, P_n implies P_{n+1}.

The principle depends for its validity on the fact that every non-empty set of natural numbers has a least element: thus if P_n were false for some n there would be a least such n, say n_0; this could not be 1 if condition (i) holds, so there would then be an n, namely $n_0 - 1$, for which P_n is true but P_{n+1} is false, contrary to condition (ii).

(a) If P_n is the proposition 'the set $E_1 \cup E_2 \cup ... \cup E_n$ is finite provided that each of the sets $E_1, E_2, ..., E_n$ is finite' then condition (i) obviously holds, and condition (ii) holds because

$$E_1 \cup E_2 \cup ... \cup E_{n+1} = (E_1 \cup E_2 \cup ... \cup E_n) \cup E_{n+1}$$

and the union of any two finite sets is finite. (If two sets have respectively m elements and n elements, their union has at most $m+n$ elements.)

(b) If P_n is the proposition

$$1^2 + 3^2 + 5^2 + ... + (2n-1)^2 = \tfrac{1}{3}(4n^3 - n)$$

then condition (i) obviously holds, and condition (ii) is easily established by elementary algebra (adding $(2n+1)^2$ to both sides of the equation that represents P_n, and then re-arranging the right-hand side).

(Mathematical induction is a particular mechanism of deduction, and is thus fundamentally different from scientific induction, which is a method of inference having an inherent possibility of error. However, when we use mathematical induction—often called 'induction' for short—we are usually seeking to prove a conjecture to which we have been led by a process of scientific induction, or by some other heuristic device which is not valid as a method of proof. Use of the imagination to produce interesting conjectures, which one subsequently tries to prove or to disprove, is an important part of creative work in mathematics.)

4. The classical proof that there are infinitely many prime numbers can be adapted in the following way. Let $p_1, ..., p_n$ be prime numbers of the form $3k+2$ (that is, congruent to 2 modulo 3). If $p_1 p_2 \ldots p_n$ is divided by 3, the remainder is 1 if n is even, 2 if n is odd. Hence either $p_1 p_2 \ldots p_n + 1$ or $p_1 p_2 \ldots p_n + 3$ is of the form $3k+2$. Now, neither of these two numbers is divisible by any of the numbers $p_1, \ldots, p_n$, but any number of the form $3k+2$ has at least one prime factor of that form (since every prime number other than 3 is either of the form $3k+1$ or the form $3k+2$, and a product of numbers of the form $3k+1$ is of the same form). Thus there is a prime number of the form $3k+2$ which is not one of the numbers $p_1, \ldots, p_n$. This shows that there are infinitely many prime numbers of the form $3k+2$.

The same type of argument can be used for prime numbers of the form $4k+3$. (Note, however, that it will not work in the case of prime numbers of the form $3k+1$, which is much more difficult.)

5. The l.c.m. of the numbers $1, \ldots, n$ can be found as follows. Express each of the numbers $2, \ldots, n$ as a product of powers of prime numbers, and for each of the distinct prime numbers so obtained choose the highest power of it that occurs in any of the factorizations: then the product of these highest prime powers is the l.c.m. (This follows from 'the fundamental theorem of arithmetic'.) Thus the l.c.m. is $2^m k$, where k is a product of powers of odd primes and is therefore odd, and 2^m is the highest power of 2 that divides any of the numbers $2, \ldots, n$. Since a factor of any of the numbers $2, \ldots, n$ is itself one of these numbers, m is the greatest integer such that $2^m \leqslant n$.

Now consider the integers $2^m k / j$, where $j = 1, 2, 3, \ldots, n$. One of these is odd (the one for which $j = 2^m$), but all the others are even: the sum of these integers is therefore an odd number, say r. If $s = 2^m k$, which is an even number, then

$$1 + \frac{1}{2} + \frac{1}{3} + \ldots + \frac{1}{n} = \frac{r}{s}.$$

In a similar way, but considering powers of 3 instead of powers of 2, we can show that, if $n > 1$,

$$1 + \frac{1}{3} + \frac{1}{5} + \ldots + \frac{1}{2n-1}$$

is the ratio of a number which is not divisible by 3 to a number which is divisible by 3.

6. We know that the h.c.f. of the integers m and n can be expressed in the form $rn + sm$, where r and s are integers (which can be calculated by Euclid's algorithm). If m and n have no common factors (other than 1) their h.c.f. is 1, so that $1 = rn + sm$; and neither m nor n is 0, so that $mn \neq 0$, and

$$\frac{1}{mn} = \frac{r}{m} + \frac{s}{n}.$$

Now any rational number can be expressed in the form

$$\frac{a}{b_1 b_2 \ldots b_n},$$

where a is an integer and $b_1, \ldots, b_n$ are powers of distinct prime numbers. We wish to show that there are integers $a_1, \ldots, a_n$ such that

$$\frac{a}{b_1 b_2 \ldots b_n} = \frac{a_1}{b_1} + \ldots + \frac{a_n}{b_n}.$$

Assume, as induction hypothesis, that this is the case for some value of n. (It is obviously so when $n = 1$.) Let b_{n+1} be a power of another prime number (possibly the first power), so that b_{n+1} has no factor in common with any of the numbers $b_1, \ldots, b_n$. Then

$$\frac{a}{b_1 b_2 \ldots b_n b_{n+1}} = \frac{a_1}{b_1 b_{n+1}} + \ldots + \frac{a_n}{b_n b_{n+1}},$$

and by the result proved above there are integers $a_1', a_1'', \ldots, a_n', a_n''$ such that

$$\frac{a_1}{b_1 b_{n+1}} = \frac{a_1'}{b_1} + \frac{a_1''}{b_{n+1}}, \quad \ldots, \quad \frac{a_n}{b_n b_{n+1}} = \frac{a_n'}{b_n} + \frac{a_n''}{b_{n+1}}.$$

Thus

$$\frac{a}{b_1 b_2 \ldots b_n b_{n+1}} = \frac{a_1'}{b_n} + \ldots + \frac{a_n'}{b_n} + \frac{a_{n+1}}{b_{n+1}},$$

where a_{n+1} is the integer $a_1'' + \ldots + a_n''$. The desired result now follows by induction.

7. For any integer m greater than 1, let M be the set of integers $0, 1, \ldots, m-1$. If $x, y \in M$ we define $x \underset{m}{+} y$ to be the remainder after division of $x + y$ by m. Thus the equation

$$x \underset{m}{+} y = z$$

has the same meaning as the congruence

$$x + y \equiv z \pmod{m}.$$

Similarly, we define $x \underset{m}{\times} y$ to be remainder after division of xy by m. It is easy to verify that with $\underset{m}{+}$ and $\underset{m}{\times}$ as 'addition' and 'multiplication' respectively, M is a commutative ring: this is *the ring of integers modulo m* (also called the ring of remainders, or of *residues*, modulo m); it has the integer 0 as its zero element, and it has a unit element, namely 1. For this ring to be a field, it is necessary and sufficient that for each non-zero element x of M there should be an element x' of M such that $x' \underset{m}{\times} x = 1$, that is, $x'x \equiv 1 \pmod{m}$. (A field is by definition a commutative ring in which every non-zero element x has a *reciprocal*, or 'multiplicative inverse'—an element whose product with x is multiplicatively neutral; see **9.**) If m is not prime then $m = xy$, where $x, y \in M$; then $x \underset{m}{\times} y = 0$, so that if $x' \underset{m}{\times} x = 1$ then

$$y = (x' \underset{m}{\times} x) \underset{m}{\times} y = x' \underset{m}{\times} (x \underset{m}{\times} y) = 0,$$

whereas $y \neq 0$ because $m \neq 0$. Thus the ring cannot be a field if m is not prime. If m is prime, let x be any non-zero element of M: then the products $0 \underset{m}{\times} x,\ 1 \underset{m}{\times} x, \ldots, (m-1) \underset{m}{\times} x$ are all different; for if $h \underset{m}{\times} x = k \underset{m}{\times} x$, where $0 \leqslant h < k < m$, then $(k-h)x \equiv 0 \pmod{m}$, and this implies, if m is prime, that m divides $k-h$ or x, both of which are less than m. The set of these products is thus the whole of M; so one of the products is 1, and it follows that the ring is a field.

(The above proof can be analysed as follows. (i) The ring in question has *divisors of zero*—that is, 0 can be expressed as a product of non-zero elements in the ring—if and only if m is composite. (ii) There can be no divisors of zero in any ring which is a field or a part of a field. (iii) A commutative ring that has a unit element and no divisors of zero—that is, an *integral domain*—is necessarily a field if it is finite. Note that an infinite integral domain need not be a field—though if it is not itself a field it can always be regarded as a part of a field, in the way that the ring of integers can be regarded as a part of the field of rational numbers.)

In the ring of integers modulo 7, 3 is the reciprocal of 5, since $3 \times 5 \equiv 1$ (mod 7). Since $2 \times 5 \equiv 1 \pmod{9}$, 5 has a reciprocal, namely 2, in the ring of integers modulo 9.

8. By the symmetry of the definition,

$$A \triangle B = B \triangle A$$

for any subsets A, B of E; the binary operation $\triangle$ is thus commutative. Also, $x \in A \triangle B$ if and only if x belongs to A or to B but not to both. Hence $x \in A \triangle (B \triangle C)$ if and only if x belongs either to only one of A, B, C, or else to all three; and we have the same criterion for x to be an element of

$(A \triangle B) \triangle C$, so the operation $\triangle$ is associative. For any subset A of E, $A \triangle \emptyset = A$, so $\emptyset$ is a neutral element for $\triangle$; and

$$A \triangle A = \emptyset,$$

so A has an inverse (namely itself) with respect to $\triangle$. Thus the subsets of E form an Abelian group with respect to $\triangle$.

The operation $\cap$ is commutative and associative. Also, if A, B, C are any subsets of E,

$$A \cap (B \triangle C) = (A \cap B) \triangle (A \cap C);$$

that is, $\cap$ is 'distributive over $\triangle$'. Thus, with $\triangle$ as addition and $\cap$ as multiplication, the subsets of E form a ring. Clearly, this ring has a unit element, namely E. If $A \cap B = E$ then $A = E$, so that E is the only element that has a reciprocal in the ring.

(We have here an example of a 'Boolean ring', so called after the nineteenth-century Anglo-Irish mathematician George Boole, whose work on 'the laws of thought' was very influential in the development of pure mathematics; see **80, 106**. The symmetric difference of two sets has sometimes been called their 'sum modulo 2'; it is the same as their union if and only if the sets are disjoint. The operation $\triangle$ is used in the subject known as 'combinatorial topology'—which is a part of 'algebraic topology'. Intersections of sets were formerly called 'products', or 'logical products'.)

9. A *field* is a commutative ring in which the non-zero elements form a group with respect to multiplication. (See **57**.)

In any ring, $$x0 = x(0+0) = x0 + x0,$$

and therefore $x0 = 0$; similarly, $0x = 0$. Hence

$$(-x)(-y) + (-x)y = (-x)(-y+y) = 0,$$

so that $(-x)(-y) = -(-x)y$; similarly,

$$(-x)y + xy = (-x+x)y = 0,$$

so that $(-x)y = -xy$. Thus $(-x)(-y) = -(-xy) = xy$. If the ring is a field, and $x \neq 0$, then x has a unique reciprocal x^{-1}: since $xx^{-1} = 1$, it follows from what has just been proved that

$$(-x)(-x^{-1}) = 1,$$

so that $-x^{-1}$ is the reciprocal of $-x$. Also, if $xy = xz$ and $x \neq 0$ then

$$y = (x^{-1}x)y = x^{-1}(xy) = x^{-1}(xz) = (x^{-1}x)z = z.$$

(Note that we have made no use here of the commutativity of multiplication, or even of the commutativity of addition; we have however used the fact that a ring is a group with respect to addition, and we have used both the left-hand distributive law and the right-hand distributive law. Thus the rules of elementary algebra that are sometimes loosely expressed by the statement 'two minuses make a plus' can be seen to be valid in a fairly general context. Up to the point where we took the ring to be a field, we made no use of the associativity of multiplication.)

10. Suppose that, in a given field, x and y are square roots of the same element. Then $x^2 = y^2$, so that

$$(x-y)(x+y) = x^2 - y^2 = 0.$$

Now, in a field a product is zero only if at least one of the factors is zero. Hence if $x \neq y$ then $x+y = 0$, so that $x = -y$. Thus no element of a field has more than two square roots in the field.

If 2 had a rational square root there would be integers p and q, with no common factor, such that
$$(p/q)^2 = 2,$$

that is, $p^2 = 2q^2$. Thus 2 would divide p^2, and would therefore divide p (since 2 is prime); but if $p = 2r$, for some integer r, then $q^2 = 2r^2$, and this implies that 2 divides q, so that p and q would have 2 as a common factor. This contradiction shows that 2 has no rational square root. (In the same way we can show that no other prime number has a rational square root.)

Since $3^2 \equiv 2 \pmod 7$, 3 is a square root of 2 in the field of integers modulo 7. (4 is another square root of 2 in this field.)

11. Since F is a field it has a unit element (with respect to the usual operation of multiplication); this is a non-zero rational number u such that $u^2 = u$, and therefore $u = 1$, the rational unit. If F contains a positive integer n then it contains $n+1$; hence, by induction, F contains every positive integer. Also, F has a zero element; this is a rational number z such that $z+1 = 1$, and therefore $z = 0$, the rational zero. It follows, since F is a field, that F contains every integer, and every quotient of an integer by a non-zero integer. Thus F consists of all the rational numbers.

(This result can be expressed by saying that the field of rational numbers has no proper subfield, or that it is a *minimal field*. For any prime number p, the field of integers modulo p is also minimal in this sense.)

12. A *totally ordered field* is a field F in which a particular set P is specified in such a way that (i) if $x \in P$ and $y \in P$ then $x + y \in P$ and $xy \in P$, and (ii) if $x \in F$ then either $x \in P$ or $-x \in P$ or x is zero, these last three possibilities being mutually exclusive. The elements of P are the *positive* elements of the totally ordered field. The conditions (i) and (ii) imply that if x is any non-zero element of F then $x^2 \in P$; so that if 1 is the unit element of F then $1 \in P$. Hence the sums $1 + 1, 1 + 1 + 1, \ldots$ belong to P. These sums are all different, for if two of them were equal then one of them would be zero and would consequently not belong to P. Hence if we write

$$1 + 1 + \ldots + 1 = n1,$$

where n is the number of terms on the left, the correspondence $n \sim n1$ is one-to-one. We can extend this correspondence by associating the rational number $(m - n)/p$ with the element $(m1 - n1)/p1$ of F, for any natural numbers m, n, p, and we thereby establish an isomorphism between the field of rational numbers and a certain subfield, Q say, of F; and the positive rational numbers then correspond to the positive elements of Q. In this sense F 'contains' the system (totally ordered field) of rational numbers; we can identify 1 with 1, $n1$ with n, and so on.

If x and y are elements of a totally ordered field, the statement 'x is greater than $y (x > y)$', or equivalently 'y is less than $x (y < x)$', means that $x - y$ is positive. Thus, in such a field, every element other than x is either less than x or greater than x. The relations $>$ and $<$ are transitive, because sums of positive elements are positive; and 'positive' has the same meaning as 'greater than zero'. (*Negative* means 'less than zero'.) The system R of real numbers is a totally ordered field with the following property: for any non-empty sets E and H such that every element of E is less than every element of H, there is at least one element ξ of R such that no element of E is greater than ξ and no element of H is less than ξ. This property distinguishes R from all other totally ordered fields (except isomorphic copies of itself); see **62, 67**. The system of rational numbers can be regarded as a part of R, but there are *irrational* real numbers also.

In a totally ordered field, any quotient of positive elements is positive: hence if we are given a positive element x we can find a positive element less than x, for instance $\frac{1}{2}x$. Now if c is a positive real number, let E be the set of all positive real numbers whose squares are less than c, and H the set of all positive real numbers whose squares are greater than c. Then E and H are non-empty ($1 \in E$ if $x > 1$, $c \in E$ if $c \geqslant 1$, and $c + 1 \in H$), and every element of E is less than every element of H. Hence there is a positive real number ξ such that no element of E is greater than ξ and no element of H is less than ξ. If $\xi^2 < c$, let δ be a positive number which is less than ξ and also less than $(c - \xi^2)/3\xi$: then

$$(\xi + \delta)^2 = \xi^2 + (2\xi + \delta)\delta < \xi^2 + 3\xi\delta < c,$$

so that $\xi + \delta \in E$, although $\xi + \delta > \xi$; this contradiction proves that $\xi^2 \geqslant c$. We can show similarly that $\xi^2 \leqslant c$, and it follows that $\xi^2 = c$.

Any subfield of R is a totally ordered field. If c is a positive rational number, the numbers of the form $r + s\sqrt{c}$, where r and s are rational, form such a subfield: this is not the whole of R, for it does not contain the square root of every positive number; and it is not the field Q of rational numbers if $\sqrt{c}$ is irrational (for example if $c = 2$). See **68**.

13. An *upper bound* of a set E of real numbers is a real number b such that no element of E is greater than b. If $E = \varnothing$ then every real number is an upper bound of E; but if $E \neq \varnothing$ and E has upper bounds then (by the special property referred to in **12**) E has a *least* upper bound (necessarily unique): this is the number $\sup E$. (The symbol 'sup' is an abbreviation of the word 'supremum', whose meaning must be carefully distinguished from that of 'maximum': the existence of $\sup E$ does not imply the existence of $\max E$, but if $\max E$ exists then $\sup E = \max E$. There is a similar distinction between 'infimum' and 'minimum'.)

If E consists of all the rational numbers less than x then x is an upper bound of E, so $\sup E \leqslant x$: if $\sup E < x$ there is a rational number r such that $\sup E < r < x$, contrary to the definition of $\sup E$; thus $\sup E = x$.

(See **62, 67**.)

14. If $x > c + 1$, where $c > 0$, then

$$x^3 > (c + 1)^3 = c^3 + 3c^2 + 3c + 1 > c,$$

so that if $x^3 \leqslant c$ then $x \leqslant c + 1$. Thus $c + 1$ is an upper bound of the set E of all real numbers x such that $x^3 \leqslant c$. Let $b = \sup E$.

There are some positive numbers in E ($1 \in E$ if $c > 1$, and $c \in E$ if $c \leqslant 1$): hence $b > 0$. If $b^3 < c$ let δ be a positive number which is less than b and also less than $(c - b^3)/7b^2$: then

$$(b + \delta)^3 = b^3 + (3b^2 + 3b\delta + \delta^2)\delta < b^3 + 7b^2\delta < c,$$

so that $b + \delta \in E$, contrary to the definition of b. Thus $b^3 \geqslant c$. If $b^3 > c$ let δ be a positive number which is less than b and also less than $(b^3 - c)/4b^2$: then, by the definition of b, there is an x in E such that $x > b - \delta$, and then

$$x^3 > (b - \delta)^3 = b^3 - (3b^2 + \delta^2)\delta + 3b\delta^2 > b^3 - (3b^2 + \delta^2)\delta > b^3 - 4b^2\delta > c;$$

but $x^3 \leqslant c$, since $c \in E$. It follows that $b^3 = c$. Hence $(-b)^3 = -c$. Thus every positive or negative real number has a real cube root; and, of course, 0 is a cube root of 0. (See **66**.)

(In fact each real number has only one real cube root, since

$$x^3 - y^3 = (x-y)(x^2 + xy + y^2) = (x-y)\{(x + \tfrac{1}{2}y)^2 + 3y^2/4\}$$

and $(x + \tfrac{1}{2}y)^2 + 3y^2/4 > 0$ if x and y are real and not both zero. However, in the larger field of complex numbers—which is not a totally ordered field—every non-zero number has three distinct cube roots.)

15. A set of real numbers is *bounded* if and only if it has upper bounds and lower bounds.

Let $b = \sup E$ and $c = \sup H$, where E and H are non-empty sets of positive real numbers. If $x \in E$ and $y \in H$ then $x \leqslant b$ and $y \leqslant c$. The first of these inequalities implies, since $y > 0$, that $xy \leqslant by$; and the second implies, since $b > 0$, that $by \leqslant bc$. Hence $xy \leqslant bc$, so that bc is an upper bound of the set EH. Thus EH is bounded (0 being a lower bound), and

$$\sup EH \leqslant bc.$$

Let $a = \sup EH$. If $a < bc$ let $\delta = (bc - a)/(b + c)$. Then $b - \delta < b$ and $c - \delta < c$, so that, for some x in E and some y in H, $x > b - \delta$ and $y > c - \delta$; then

$$xy > (b - \delta)(c - \delta) = bc - (b + c)\delta + \delta^2 > bc - (b + c)\delta = a,$$

whereas $xy \leqslant a$, since $xy \in EH$. It follows that $a = bc$.

If E consists of the numbers 0 and -1, and if $H = E$, then

$$\sup E = \sup H = 0$$

but $\sup EH = 1$, so that in this case $\sup EH \neq \sup E \sup H$.

16. If $0 \leqslant a < b$ in a totally ordered field then, as is easy to prove, $a \leqslant x < b$ if and only if $a^2 \leqslant x^2 < b^2$.

Now the statement that $1 \cdot 022$ is an accurate estimate of x to 3 decimal places means that
$$1 \cdot 0215 \leqslant x < 1 \cdot 0225,$$

and hence that $\quad 1 \cdot 04346225 \leqslant x^2 < 1 \cdot 04550625,$

which does not imply that $1 \cdot 0435 \leqslant x^2 < 1 \cdot 0445$. Thus we cannot infer, from the given information about x, that $1 \cdot 044$ is an accurate estimate of x^2 to 3 decimal places. However, since

$$(1 \cdot 0105)^2 < 1 \cdot 0215 \quad \text{and} \quad 1 \cdot 0225 < (1 \cdot 0115)^2,$$

the given information does imply that

$$1 \cdot 0105 \leqslant \sqrt{x} < 1 \cdot 0115,$$

which is to say that $1 \cdot 011$ is an accurate estimate of $\sqrt{x}$ to 3 decimal places.

17. Since $k > 1$, $\inf E > 0$ if and only if

$$k \inf E > \inf E.$$

But, by the definition of $\inf E$, $a > \inf E$ if and only if there is an x in E such that $x < a$. Thus $\inf E > 0$ if and only if there is an x in E such that $x < k \inf E$.

Since E_1 and E_2 are sets of positive numbers, $\inf E_1 \geqslant 0$ and $\inf E_2 \geqslant 0$. If $\inf E_1 > 0$ there is an element k^{-n} of E_1 such that

$$k^{-n} < k \inf E_1,$$

so that $k^{-n-1} < \inf E_1$, which is impossible since $k^{-n-1} \in E_1$. Thus $\inf E_1 = 0$. To deal with E_2, let E be the set of all numbers $k^n/n!$ with $n > k^2 - 1$; then $E \subseteq E_2$, and therefore $\inf E \geqslant \inf E_2$. If $\inf E > 0$ there is an n greater than $k^2 - 1$ such that

$$\frac{k^n}{n!} < k \inf E$$

and consequently $\quad \dfrac{k^{n+1}}{(n+1)!} < \dfrac{k^2}{n+1} \inf E < \inf E,$

which is impossible since $k^{n+1}/(n+1)! \in E$. Thus $\inf E = 0$, and therefore $\inf E_2 = 0$.

(If $0 < k < 1$ then $\inf E_1 = k^{-1}$, but $\inf E_2 = 0$. What if $k < 0$?)

18. Let $f(x) = ax^2 + 2bx + c$, where a, b, c are real, and suppose that $f(x) \geqslant 0$ for every real x. If $a = 0$ then $b = 0$, for otherwise we could take x to be $-(b^2 + c)/2b$, making $f(x)$ negative: thus if $a = 0$ then $b^2 = ac$. If $a > 0$ then $\quad af(-b/a) = ac - b^2,$

so $b^2 \leqslant ac$. The case in which $a < 0$ does not arise, because, if $a \neq 0$,

$$f(x) = a \left\{ \left(x + \frac{b}{a} \right)^2 + \frac{c}{a} - \frac{b^2}{a^2} \right\},$$

which has the same sign as a when $|x|$ is sufficiently large.

Now let
$$a = a_1^2 + \ldots + a_n^2,$$
$$b = a_1 b_1 + \ldots + a_n b_n,$$
$$c = b_1^2 + \ldots + b_n^2.$$

Then $$f(x) = (a_1 x + b_1)^2 + \dots + (a_n x + b_n)^2,$$

so that $f(x) \geqslant 0$ for all real values of x. Hence $b^2 \leqslant ac$. Therefore

$$(a_1 + b_1)^2 + \dots + (a_n + b_n)^2 = a + 2b + c$$
$$\leqslant a + 2(ac)^{\frac{1}{2}} + c = (a^{\frac{1}{2}} + c^{\frac{1}{2}})^2,$$

so that

$$\{(a_1 + b_1)^2 + \dots + (a_n + b_n)^2\}^{\frac{1}{2}} \leqslant (a_1^2 + \dots + a_n^2)^{\frac{1}{2}} + (b_1^2 + \dots + b_n^2)^{\frac{1}{2}}.$$

(Note that the last inequality is strict unless $b^2 = ac$, and that if $a_1, \dots, a_n$ are not all zero then $b^2 = ac$ only if there is a real number x for which $f(x) = 0$, that is, only if the n simultaneous equations

$$a_1 x + b_1 = 0, \quad \dots, \quad a_n x + b_n = 0$$

have a solution. Note too that we can reverse the signs of all the numbers $b_1, \dots, b_n$—a change which does not affect the right-hand side; and that the last inequality can then be interpreted geometrically as stating that the length of any side of a triangle in n-dimensional Euclidean space is less than the sum of the lengths of the other two sides—hence the name 'triangle inequality'. The inequality

$$(a_1 b_1 + \dots + a_n b_n)^2 \leqslant (a_1^2 + \dots + a_n^2)(b_1^2 + \dots + b_n^2)$$

can be interpreted in terms of 'scalar products'—see **72**; it has various generalizations, and is associated with the names of the nineteenth-century mathematicians Cauchy, Schwarz, and Bunjakowski—respectively French, German, and Russian.)

19. A *complex number* is an ordered pair, (x, y), of real numbers; that is to say, it is a 2-dimensional real vector. Complex numbers are added and multiplied according to the following definitions of sum and product:

$$(x_1, y_1) + (x_2, y_2) = (x_1 + x_2, \; y_1 + y_2),$$
$$(x_1, y_1)(x_2, y_2) = (x_1 x_2 - y_1 y_2, \; x_1 y_2 + x_2 y_1).$$

With these definitions the set of all complex numbers becomes a field, in which the elements of the form $(x, 0)$ constitute a subfield isomorphic, under the correspondence $(x, 0) \sim x$, with the field of real numbers. It is customary to identify $(x, 0)$ with x, and to denote $(0, 1)$ by i, so that (x, y) may be written as $x + iy$. If we write z for $x + iy$, we often write $\operatorname{re} z$ for x and $\operatorname{im} z$ for y, and call these real numbers the *real part* and the *imaginary part* of z respectively; $|z|$, the *modulus* of z, is the non-negative real number $(x^2 + y^2)^{\frac{1}{2}}$. The complex numbers i and $-i$ are square roots of

-1; and i is sometimes written as $\sqrt{(-1)}$. (For reasons of their own, electrical engineers often denote $\sqrt{(-1)}$ by j rather than i.)

A square root of (x,y) is a complex number (ξ,η) such that

$$(\xi^2-\eta^2,\ 2\xi\eta) = (x,y),$$

that is, such that $\xi^2-\eta^2 = x$ and $2\xi\eta = y$.

If $y = 0$ and $x \geqslant 0$ these equations are satisfied if $\eta = 0$ and $\xi = x^{\frac12}$; if $y = 0$ and $x < 0$ they are satisfied if $\xi = 0$ and $\eta = (-x)^{\frac12}$. If $y \neq 0$ they are satisfied if $\eta = y/2\xi$ and

$$\xi^2-\frac{y^2}{4\xi^2} = x;$$

and this last equation is equivalent to the equation

$$(\xi^2-\tfrac12 x)^2 = \tfrac14(x^2+y^2),$$

which is satisfied if ξ is a square root of the positive number

$$\tfrac12\{x+(x^2+y^2)^{\frac12}\}.$$

Thus (x,y) has a square root in every case. (Alternatively, we can observe that $(z+|z|)^2 = \rho z$, where $\rho = 2(|z|+\mathrm{re}\,z)$; if z is not real then $\rho > 0$ and hence $\rho^{-\frac12}(z+|z|)$ is a square root of z. If z is real and negative then $i(-z)^{\frac12}$ is a square root of z.)

Any quadratic equation in the field of complex numbers can be solved by the method of 'completing the square', which succeeds because of the existence of square roots in all cases.

(The field of complex numbers has a much more remarkable property, discovered at the end of the eighteenth century by the German mathematician C. F. Gauss: it is an *algebraically closed* field, in the sense that every polynomial equation whose coefficients are in the field has a root in the field. The theorem that tells us this is known as *the fundamental theorem of algebra*, though it is essentially a theorem of analysis. Algebraic methods—methods involving only rational processes and 'root-extraction'—are available for solving all cubic and quartic equations in the field of complex numbers, but it is known that no such method can exist for the general equation of degree n if $n > 4$. See **71, 85**.)

20. It is easy to verify that, with the given operations, $F \times F$ is a commutative ring in which the field F is isomorphically embedded by the correspondence $x \sim (x,0)$; and this is true for any field F. If (and only if) F contains no square root of -1, then $x^2+y^2 = 0$ in F if and only if $x = 0$

and $y = 0$, so that any non-zero element (x, y) of the ring $F \times F$ has a reciprocal in $F \times F$, namely

$$(x/(x^2 + y^2), -y/(x^2 + y^2)),$$

and so $F \times F$ is a field. If $z = (0, 1)$, or if $z = (0, -1)$, then $z^2 = (-1, 0)$ so that $z^2 + 1 = 0$. (The interesting point here is that the method of constructing the field of complex numbers from that of the real numbers can be applied to any field in which there is no square root of -1, giving a larger field in which there is a square root of -1. Similar constructions can be used to 'create' other square roots. See **71**.)

If p is a prime number, the field of integers modulo p contains a square root of -1 if, for example, $p - 1$ is a perfect square. But there are prime numbers p (3 and 7 being among them) such that if F is the field of integers modulo p then $F \times F$, with the given operations, is a field having p^2 elements.

21. For any complex number z, the *conjugate* number z^* (sometimes denoted by $\bar{z}$) is defined to be $\operatorname{re} z - i \operatorname{im} z$. Thus $\operatorname{re} z^* = \operatorname{re} z$ and $\operatorname{im} z^* = -\operatorname{im} z$. It follows at once that $z^{**} = z$ for every complex number z, and that the real numbers are the self-conjugate complex numbers. (Conjugacy can thus be interpreted geometrically as a symmetry of the plane of complex numbers about the 'real axis'.)

The correspondence $z \sim z^*$ is one-to-one, and if $\alpha = x + iy$ and $\beta = u + iv$, where x, y, u, v are real, then

$$(\alpha + \beta)^* = x + u - i(y + v) = x - iy + u - iv = \alpha^* + \beta^*,$$

$$(\alpha\beta)^* = xu - yv - i(xv + yu) = (x - iy)(u - iv) = \alpha^*\beta^*.$$

The correspondence is thus an isomorphism of the field of complex numbers with itself (it is an *automorphism* of the field). Hence if

$$f(\lambda) = a_0 \lambda^n + a_1 \lambda^{n-1} + \ldots + a_{n-1} \lambda + a_n,$$

then
$$f(\lambda^*) = a_0(\lambda^*)^n + a_1(\lambda^*)^{n-1} + \ldots + a_{n-1} \lambda^* + a_n$$

$$= a_0(\lambda^n)^* + a_1(\lambda^{n-1})^* + \ldots + a_{n-1} \lambda^* + a_n,$$

$$f(\lambda)^* = a_0^*(\lambda^n)^* + a_1^*(\lambda^{n-1})^* + \ldots + a_{n-1}^* \lambda^* + a_n^*.$$

It follows that if $a_0, a_1, \ldots, a_n$ are all real then $f(\lambda^*) = f(\lambda)^*$, so that, in this case, $f(\lambda^*) = 0$ if (and only if) $f(\lambda) = 0$.

(Another property of the automorphism $z \sim z^*$ is that $|z^*| = |z|$ for every complex number z. It is interesting to note that there are only two 'modulus-preserving' automorphisms of the complex field, the other being the one in which every number corresponds to itself. To prove this, observe first that for any automorphism $z \sim z'$ of the field, $1' = 1$ and hence $r' = r$ for every rational number r. Then if x is any real number, and r is rational,

$$|x' - x| \leqslant |x' - r| + |r - x| = |(x - r)'| + |x - r|,$$

so that if moduli are preserved then $|x' - x| \leqslant 2|x - r|$; and since r can be chosen as close to x as we please, it follows that $|x' - x| = 0$, which is to say that $x' = x$. Finally, $(i')^2 = (i^2)' = (-1)' = -1$, so that $i' = \pm i$. See **71, 88**. We have here an example of an 'isometry': see **73**.)

22. A 3-*dimensional real vector* is an ordered triad, (x, y, z), of real numbers (*scalars*). If $\mathbf{r}_1, ..., \mathbf{r}_n$ are n such vectors, say

$$\mathbf{r}_1 = (x_1, y_1, z_1), \quad ..., \quad \mathbf{r}_n = (x_n, y_n, z_n),$$

and if $\alpha_1, ..., \alpha_n$ are n scalars the *linear combination* $\alpha_1 \mathbf{r}_1 + ... + \alpha_n \mathbf{r}_n$ is, by definition, the vector (x, y, z) for which

$$x = \alpha_1 x_1 + ... + \alpha_n x_n,$$
$$y = \alpha_1 y_1 + ... + \alpha_n y_n,$$
$$z = \alpha_1 z_1 + ... + \alpha_n z_n;$$

in abbreviated notation,

$$\sum_\nu \alpha_\nu \mathbf{r}_\nu = (\sum_\nu \alpha_\nu x_\nu, \sum_\nu \alpha_\nu y_\nu, \sum_\nu \alpha_\nu z_\nu).$$

If E consists of all the linear combinations of $\mathbf{r}_1, ..., \mathbf{r}_n$, let $\mathbf{v}_1, ..., \mathbf{v}_m$ be m elements of E. Then there are mn scalars

$$\alpha_{1,1}, \ ..., \ \alpha_{1,n}, \ ..., \ \alpha_{m,1}, \ ..., \ \alpha_{m,n}$$

such that $\mathbf{v}_1 = \sum_\nu \alpha_{1,\nu} \mathbf{r}_\nu, \ ..., \ \mathbf{v}_m = \sum_\nu \alpha_{m,\nu} \mathbf{r}_\nu$. Hence, for any scalars $\beta_1, ..., \beta_m$,

$$\beta_1 \mathbf{v}_1 + ... + \beta_m \mathbf{v}_m = \sum_\nu \gamma_\nu \mathbf{r}_\nu,$$

where $\gamma_\nu = \alpha_{1,\nu} \beta_1 + ... + \alpha_{m,\nu} \beta_m \ (\nu = 1, ..., n)$. Thus any linear combination of elements of E is itself an element of E.

Suppose that $\mathbf{p}$, $\mathbf{q}$, and $\mathbf{r}$ are linear combinations of $\mathbf{r}_1$ and $\mathbf{r}_2$. If $\mathbf{p} = \mathbf{0}$ then $\mathbf{p} = 0\mathbf{q} + 0\mathbf{r}$, so suppose that $\mathbf{p} \neq \mathbf{0}$. Then $\mathbf{p} = \alpha_1 \mathbf{r}_1 + \alpha_2 \mathbf{r}_2$ and the scalars α_1, α_2 are not both zero; suppose that $\alpha_1 \neq 0$. Then $\mathbf{r}_1$ is a linear combination of $\mathbf{p}$ and $\mathbf{r}_2$ (in fact $\mathbf{r}_1 = \alpha_1^{-1} \mathbf{p} - \alpha_1^{-1} \alpha_2 \mathbf{r}_2$); so $\mathbf{q}$, being a linear combination of $\mathbf{r}_1$ and $\mathbf{r}_2$, is a linear combination of $\mathbf{p}$ and $\mathbf{r}_2$: we can write $\mathbf{q} = \beta_1 \mathbf{p} + \beta_2 \mathbf{r}_2$. If $\beta_2 = 0$ then $\mathbf{q} = \beta_1 \mathbf{p} + 0\mathbf{r}$, so suppose that $\beta_2 \neq 0$. Then $\mathbf{r}_2$ is a linear combination of $\mathbf{p}$ and $\mathbf{q}$, and hence $\mathbf{r}_1$ also is a linear combination of $\mathbf{p}$ and $\mathbf{q}$ (since it is a linear combination of $\mathbf{p}$ and $\mathbf{r}_2$). It follows that $\mathbf{r}$ is a linear combination of $\mathbf{p}$ and $\mathbf{q}$.

(Note that in this discussion we have used the fact that the system of real numbers is a field, but we have used no other properties of this system. Accordingly, the whole discussion is valid for 'vectors over a field F', provided that the term 'scalar' is understood to mean 'element of F'. Moreover, the vectors need not be 3-dimensional: the number 3 can be replaced here by any natural number, provided that this 'dimension' is the same for all the vectors considered. The discussion can be extended, in an obvious way, to provide a proof of the basis theorem. See **119.**)

23. If $r_1 = (x_1, y_1, z_1)$ and $r_2 = (x_2, y_2, z_2)$ then, by definition,

$$r_1 . r_2 = x_1 x_2 + y_1 y_2 + z_1 z_2,$$

and $r_1 . r_2$ is called the *scalar product* of the vectors r_1 and r_2 (evidently, $r_1 . r_2 = r_2 . r_1$). The *length*, $|r|$, of a real vector r is the non-negative scalar $\sqrt{(r . r)}$. The statement that the vectors r_1 and r_2 are *orthogonal* to each other means that $r_1 . r_2 = 0$.

If $p = (1, 0, 2)$ and $q = (2, 2, -1)$ then

$$p . q = 2 + 0 - 2 = 0,$$

so that p and q are orthogonal to each other. A vector (x, y, z) is orthogonal to both p and q if

$$x + 2z = 0 \quad \text{and} \quad 2x + 2y - z = 0,$$

and these equations are satisfied if $x = -2z$ and $y = \frac{5}{2}z$, z being arbitrary. Now if $r = (-2z, \frac{5}{2}z, z)$ then

$$|r| = |z| \, |(-2, \tfrac{5}{2}, 1)| = |z| \, \tfrac{3}{2} \sqrt{5} = \tfrac{3}{2} |z| \, |p|$$

so that $|r| = |p|$ if $z = \pm \frac{2}{3}$. Thus each of the vectors $\pm(-\frac{4}{3}, \frac{5}{3}, \frac{2}{3})$ has the same length as p and is orthogonal to both p and q.

(A general method of finding a vector r orthogonal to each of two 3-dimensional vectors p and q is to calculate the vector product $p \wedge q$ and then take r to be $\alpha(p \wedge q)$ for any scalar α. If $p \wedge q \neq 0$, α can be chosen so that r has a prescribed length. Note that in defining the length of a real vector we use the fact that the system of real numbers is a totally ordered field in which every positive element has a square root: we cannot assign a 'length' in this way to a vector over an arbitrary field.)

24. If $r_1 = (x_1, y_1, z_1)$ and $r_2 = (x_2, y_2, z_2)$ then, by definition,

$$r_1 \wedge r_2 = (y_1 z_2 - y_2 z_1, z_1 x_2 - z_2 x_1, x_1 y_2 - x_2 y_1).$$

Thus

$$r_2 \wedge r_1 = (y_2 z_1 - y_1 z_2, z_2 x_1 - z_1 x_2, x_2 y_1 - x_1 y_2) = -r_1 \wedge r_2.$$

Also,

$$\mathbf{r}_1 . \mathbf{r}_1 \wedge \mathbf{r}_2 = x_1(y_1 z_2 - y_2 z_1) + y_1(z_1 x_2 - z_2 x_1) + z_1(x_1 y_2 - x_2 y_1) = 0,$$

and $\mathbf{r}_2 . \mathbf{r}_1 \wedge \mathbf{r}_2 = -\mathbf{r}_2 . \mathbf{r}_2 \wedge \mathbf{r}_1 = 0$, so that $\mathbf{r}_1 \wedge \mathbf{r}_2$ is orthogonal to $\mathbf{r}_1$ and to $\mathbf{r}_2$.

If $\mathbf{r}_1 . \mathbf{r}_2 \wedge \mathbf{r}_3 = 1$ and $\mathbf{r} = \alpha_1 \mathbf{r}_1 + \alpha_2 \mathbf{r}_2 + \alpha_3 \mathbf{r}_3$

then $\mathbf{r} . \mathbf{r}_2 \wedge \mathbf{r}_3 = \alpha_1 \mathbf{r}_1 . \mathbf{r}_2 \wedge \mathbf{r}_3 + \alpha_2 \mathbf{r}_2 . \mathbf{r}_2 \wedge \mathbf{r}_3 + \alpha_3 \mathbf{r}_3 . \mathbf{r}_2 \wedge \mathbf{r}_3 = \alpha_1,$

since $\mathbf{r}_2 \wedge \mathbf{r}_3$ is orthogonal to $\mathbf{r}_2$ and to $\mathbf{r}_3$. Similarly, we find that

$$\alpha_2 = \mathbf{r} . \mathbf{r}_3 \wedge \mathbf{r}_1 \quad \text{and} \quad \alpha_3 = \mathbf{r} . \mathbf{r}_1 \wedge \mathbf{r}_2.$$

(We write $\mathbf{r}_1 . \mathbf{r}_2 \wedge \mathbf{r}_3$ for $\mathbf{r}_1 . (\mathbf{r}_2 \wedge \mathbf{r}_3)$ since there is no risk of ambiguity; and since $\mathbf{r}_1 . \mathbf{r}_2 \wedge \mathbf{r}_3 = \mathbf{r}_1 \wedge \mathbf{r}_2 . \mathbf{r}_3$ this *scalar triple product* is sometimes denoted by $[\mathbf{r}_1, \mathbf{r}_2, \mathbf{r}_3]$. Note that if any two of the factors in a scalar triple product are interchanged then the sign of the product is reversed. Note also that although the vectors involved in this discussion must be 3-dimensional, they need not be real: the whole discussion is valid for 3-dimensional vectors over any field. The theory of scalar triple products is essentially the theory of 3-by-3 determinants. See **118**.)

25. If α and β are any scalars then

$$(\alpha \mathbf{u} + \beta \mathbf{v}) . \mathbf{u} \wedge \mathbf{v} = \alpha \mathbf{u} . \mathbf{u} \wedge \mathbf{v} + \beta \mathbf{v} . \mathbf{u} \wedge \mathbf{v};$$

but this is 0 because $\mathbf{u} \wedge \mathbf{v}$ is orthogonal to $\mathbf{u}$ and to $\mathbf{v}$. Thus $\mathbf{u} \wedge \mathbf{v}$ is orthogonal to every linear combination of $\mathbf{u}$ and $\mathbf{v}$. Hence if

$$\mathbf{r} = \alpha \mathbf{u} + \beta \mathbf{v} + \gamma \mathbf{u} \wedge \mathbf{v},$$

then $\mathbf{r} . \mathbf{u} \wedge \mathbf{v} = \gamma |\mathbf{u} \wedge \mathbf{v}|^2.$

Now suppose that $\mathbf{u} \wedge \mathbf{v} \neq \mathbf{0}$ (so that $|\mathbf{u} \wedge \mathbf{v}| \neq 0$) and that $\mathbf{r} = \mathbf{0}$. Then $\gamma = 0$. Hence

$$\mathbf{0} = \alpha \mathbf{u} + \beta \mathbf{v},$$

and therefore (since $\mathbf{u} \wedge \mathbf{0} = \mathbf{0}$)

$$\mathbf{0} = \alpha \mathbf{u} \wedge \mathbf{u} + \beta \mathbf{u} \wedge \mathbf{v} = \beta \mathbf{u} \wedge \mathbf{v},$$

so that $\beta = 0$. Therefore $\mathbf{0} = \alpha \mathbf{u}$, so that $\alpha = 0$ ($\mathbf{u} \neq \mathbf{0}$ since $\mathbf{u} \wedge \mathbf{v} \neq \mathbf{0}$). Thus, if $\mathbf{u} \wedge \mathbf{v} \neq \mathbf{0}$, the equation $\mathbf{r} = \mathbf{0}$ is satisfied only if the scalars α, β, and γ are all zero; and this means that the vectors $\mathbf{u}$, $\mathbf{v}$, and $\mathbf{u} \wedge \mathbf{v}$ are linearly independent.

It is a consequence of the basis theorem, for 3-dimensional vectors, that no 4 vectors can be linearly independent (see **119**). Thus for any

4 vectors $\mathbf{u}$, $\mathbf{v}$, $\mathbf{w}$, $\mathbf{r}$ there are scalars, α, β, γ, δ, not all zero, such that

$$\alpha\mathbf{u} + \beta\mathbf{v} + \gamma\mathbf{w} + \delta\mathbf{r} = \mathbf{0}.$$

If $\mathbf{u}$, $\mathbf{v}$, and $\mathbf{w}$ are linearly independent then $\delta \neq 0$ (whatever $\mathbf{r}$ may be), so that $\mathbf{r}$ is a linear combination of $\mathbf{u}$, $\mathbf{v}$, and $\mathbf{w}$. In particular, if $\mathbf{u} \wedge \mathbf{v} \neq \mathbf{0}$ then any vector $\mathbf{r}$ is a linear combination of $\mathbf{u}$, $\mathbf{v}$, and $\mathbf{u} \wedge \mathbf{v}$. But we have seen that if
$$\mathbf{r} = \alpha\mathbf{u} + \beta\mathbf{v} + \gamma\mathbf{u} \wedge \mathbf{v}$$

then $\mathbf{r} . \mathbf{u} \wedge \mathbf{v} = \gamma |\mathbf{u} \wedge \mathbf{v}|^2$; thus if $\mathbf{r}$ is orthogonal to $\mathbf{u} \wedge \mathbf{v}$, and $\mathbf{u} \wedge \mathbf{v} \neq \mathbf{0}$, then $\gamma = 0$, so that $\mathbf{r}$ is a linear combination of $\mathbf{u}$ and $\mathbf{v}$.

(Note that we have proved that if $\mathbf{u} \wedge \mathbf{v} \neq \mathbf{0}$ then a vector is a linear combination of $\mathbf{u}$ and $\mathbf{v}$ if and only if it is orthogonal to $\mathbf{u} \wedge \mathbf{v}$; and that the whole discussion is valid for 3-dimensional vectors over any field.)

26. The vectors $\mathbf{r}$ for which $\mathbf{r}^2 = 1$ (the *unit vectors*) constitute the sphere with centre $\mathbf{0}$ and radius 1. Those for which $\mathbf{r} . \mathbf{e} = 1$ constitute the plane through $\frac{1}{3}\mathbf{e}$ orthogonal to $\mathbf{e}$. The set C is the intersection of these two surfaces. Since
$$(\mathbf{r} - \tfrac{1}{3}\mathbf{e})^2 = \mathbf{r}^2 - \tfrac{2}{3}\mathbf{r} . \mathbf{e} + \tfrac{1}{3},$$

$\mathbf{r}$ lies on C if and only if $\mathbf{r} . \mathbf{e} = 1$ and

$$(\mathbf{r} - \tfrac{1}{3}\mathbf{e})^2 = \tfrac{2}{3}.$$

(Thus C is a circle, with centre $\frac{1}{3}\mathbf{e}$ and radius $\sqrt{\frac{2}{3}}$.) We can choose non-zero vectors $\mathbf{f}$ and $\mathbf{g}$, of prescribed length λ say, which are orthogonal to each other and to $\mathbf{e}$. Then $\mathbf{r} . \mathbf{e} = 1$ if and only if

$$\mathbf{r} = \tfrac{1}{3}\mathbf{e} + x\mathbf{f} + y\mathbf{g},$$

where x and y are real numbers. Since $\mathbf{e}$, $\mathbf{f}$, and $\mathbf{g}$ are linearly independent, this last equation establishes a one-to-one correspondence $r \sim (x, y)$, between the plane $\mathbf{r} . \mathbf{e} = 1$ and the complex plane. Since

$$(\mathbf{r} - \tfrac{1}{3}\mathbf{e})^2 = (x^2 + y^2)\lambda^2,$$

the points of C are represented, in this correspondence, by the complex numbers (x, y) for which
$$(x^2 + y^2)\lambda^2 = \tfrac{2}{3}.$$

Thus if we take λ to be $\sqrt{\frac{2}{3}}$ we obtain a one-to-one correspondence between C and the unit circle of complex numbers.

27. According to the given definition of multiplication,

$$((\mathbf{r}_1, t_1)(\mathbf{r}_2, t_2))(\mathbf{r}_3, t_3) = (\mathbf{r}_1 \wedge \mathbf{r}_2 + t_1 \mathbf{r}_2 + t_2 \mathbf{r}_1, t_1 t_2 - \mathbf{r}_1 . \mathbf{r}_2)(\mathbf{r}_3, t_3)$$

$$= (\mathbf{r}, t),$$

where

$$\mathbf{r} = (\mathbf{r}_1 \wedge \mathbf{r}_2) \wedge \mathbf{r}_3 + t_1 \mathbf{r}_2 \wedge \mathbf{r}_3 + t_2 \mathbf{r}_1 \wedge \mathbf{r}_3 + (t_1 t_2 - \mathbf{r}_1 . \mathbf{r}_2) \mathbf{r}_3$$
$$+ t_3 (\mathbf{r}_1 \wedge \mathbf{r}_2 + t_1 \mathbf{r}_2 + t_2 \mathbf{r}_1),$$

$$t = t_1 t_2 t_3 - \mathbf{r}_1 . \mathbf{r}_2 t_3 - \mathbf{r}_1 \wedge \mathbf{r}_2 . \mathbf{r}_3 - t_1 \mathbf{r}_2 . \mathbf{r}_3 - t_2 \mathbf{r}_1 . \mathbf{r}_3;$$

and

$$(\mathbf{r}_1 t_1) ((\mathbf{r}_2, t_2) (\mathbf{r}_3, t_3)) = (\mathbf{r}_1, t_1) (\mathbf{r}_2 \wedge \mathbf{r}_3 + t_2 \mathbf{r}_3 + t_3 \mathbf{r}_2, t_2 t_3 - \mathbf{r}_2 . \mathbf{r}_3)$$
$$= (\mathbf{r}', t'),$$

where

$$\mathbf{r}' = \mathbf{r}_1 \wedge (\mathbf{r}_2 \wedge \mathbf{r}_3) + t_2 \mathbf{r}_1 \wedge \mathbf{r}_3 + t_3 \mathbf{r}_1 \wedge \mathbf{r}_2 + t_1 (\mathbf{r}_2 \wedge \mathbf{r}_3 + t_2 \mathbf{r}_3 + t_3 \mathbf{r}_2)$$
$$+ (t_2 t_3 - \mathbf{r}_2 . \mathbf{r}_3) \mathbf{r}_1,$$

$$t' = t_1 t_2 t_3 - t_1 \mathbf{r}_2 . \mathbf{r}_3 - \mathbf{r}_1 . \mathbf{r}_2 \wedge \mathbf{r}_3 - t_2 \mathbf{r}_3 . \mathbf{r}_1 - t_3 \mathbf{r}_1 . \mathbf{r}_2.$$

Now

$$\mathbf{r} - \mathbf{r}' = (\mathbf{r}_1 \wedge \mathbf{r}_2) \wedge \mathbf{r}_3 - \mathbf{r}_1 \wedge (\mathbf{r}_2 \wedge \mathbf{r}_3) - \mathbf{r}_1 . \mathbf{r}_2 \mathbf{r}_3 + \mathbf{r}_2 . \mathbf{r}_3 \mathbf{r}_1 = \mathbf{0}$$

by the given identity, so $\mathbf{r} = \mathbf{r}'$; and $t = t'$ because $\mathbf{r}_1 \wedge \mathbf{r}_2 . \mathbf{r}_3 = \mathbf{r}_1 . \mathbf{r}_2 \wedge \mathbf{r}_3$. Thus $(\mathbf{r}, t) = (\mathbf{r}', t')$, so that multiplication is associative. It is not commutative, since

$$(\mathbf{r}_1, 0) (\mathbf{r}_2, 0) = (\mathbf{r}_1 \wedge \mathbf{r}_2, -\mathbf{r}_1 . \mathbf{r}_2), \quad (\mathbf{r}_2, 0) (\mathbf{r}_1, 0) = (\mathbf{r}_2 \wedge \mathbf{r}_1, -\mathbf{r}_2 . \mathbf{r}_1),$$

and

$$\mathbf{r}_1 \wedge \mathbf{r}_2 \neq \mathbf{r}_2 \wedge \mathbf{r}_1 \quad \text{if} \quad \mathbf{r}_1 \wedge \mathbf{r}_2 \neq \mathbf{0}.$$

For any vector $\mathbf{r}$ and any scalar t,

$$(\mathbf{0}, 1) (\mathbf{r}, t) = (\mathbf{r}, t) \quad \text{and} \quad (\mathbf{r}, t) (\mathbf{0}, 1) = (\mathbf{r}, t),$$

by the laws of vector algebra. Also,

$$(\mathbf{r}, t) (-\mathbf{r}, t) = (-\mathbf{r} \wedge \mathbf{r}, t^2 + \mathbf{r} . \mathbf{r}) = (\mathbf{0}, t^2 + \mathbf{r}^2) = (-\mathbf{r}, t) (\mathbf{r}, t).$$

If $(\mathbf{r}, t) \neq (\mathbf{0}, 0)$ then $t^2 + \mathbf{r}^2$ is a positive real number; and if

$$\mathbf{s} = -(t^2 + \mathbf{r}^2)^{-1} \mathbf{r} \quad \text{and} \quad u = (t^2 + \mathbf{r}^2)^{-1}$$

then

$$(\mathbf{s}, u) (\mathbf{r}, t) = (\mathbf{r}, t) (\mathbf{s}, u) = (\mathbf{0}, 1).$$

(With multiplication defined in this way, and with addition defined in the usual way, the 4-dimensional real vectors form a ring; this is a *division ring*, in that every non-zero element has a two-sided multiplicative inverse. The elements of this ring are known as *quaternions*. Those of the type $(\mathbf{0}, t)$ form a subring isomorphic with the field of real numbers; and there are subrings isomorphic with the field of complex numbers. If $\mathbf{i}, \mathbf{j}, \mathbf{k}$ is a right-handed orthonormal sequence, the four quaternions $(\mathbf{i}, 0)$, $(\mathbf{j}, 0)$, $(\mathbf{k}, 0)$, $(0, 1)$ have an interesting multiplication table. If the n-dimensional real vectors, with addition defined in the usual way, are given a scheme of multiplication so as to form a division ring R containing, in a certain natural sense, the field of real numbers, then by a famous theorem due to the nineteenth-century German mathematician F. G. Frobenius, either $n = 1$ and R is the field of real numbers, or $n = 2$ and R is the field of complex numbers, or $n = 4$ and R is the ring of quaternions: there are no other possibilities. Quaternions were introduced—though not in the way described here—by the nineteenth-century Irish mathematician W. R. Hamilton, who wrote extensively on their uses in mathematical physics.)

28. To find the h.c.f. of polynomials $p(x)$ and $q(x)$ we can use Euclid's algorithm, to give a succession of remainders, of decreasing degrees, each of which is divisible by every polynomial that divides both $p(x)$ and $q(x)$: the last non-zero remainder divides both $p(x)$ and $q(x)$, and is therefore the h.c.f. In particular,

$$x^4 + 2x^3 + x^2 - 1 = x^4 + x^2 + 1 + 2(x^3 - 1),$$

$$x^4 + x^2 + 1 = x(x^3 - 1) + x^2 + x + 1,$$

$$x^3 - 1 = (x - 1)(x^2 + x + 1),$$

so that the h.c.f. of the given polynomials is $x^2 + x + 1$, unless the field of coefficients is such that $2 = 0$ (in which case the given polynomials are equal). As a check, note that

$$x^4 + x^2 + 1 = (x^2 + x + 1)(x^2 - x + 1),$$

$$x^4 + 2x^3 + x^2 - 1 = (x^2 + x + 1)(x^2 + x - 1),$$

$$2(x^2 + x + 1) = (x + 2)(x^4 + x^2 + 1) - x(x^4 + 2x^3 + x^2 - 1).$$

29. If $q > 0$ then

$$q^n |f(p/q)| = q^n |a_0(p/q)^n + a_1(p/q)^{n-1} + \ldots + a_{n-1}p/q + a_n|$$

$$= |a_0 p^n + a_1 p^{n-1} q + \ldots + a_{n-1} p q^{n-1} + a_n q^n|,$$

and this is an integer if p and q and the coefficients $a_0, a_1, \ldots, a_n$ are integers. Moreover, this integer is positive unless $f(p/q) = 0$, so if $f(p/q) \neq 0$ then $q^n |f(p/q)| \geq 1$, that is
$$|f(p/q)| \geq 1/q^n.$$

By the remainder theorem, if $f(\lambda) = 0$ then, for any number x,

$$f(x) = (x - \lambda)(b_0 x^{n-1} + b_1 x^{n-1} + \ldots + b_{n-1}),$$

where the coefficients b_0, b_1, ..., b_{n-1} do not depend on x. Now unless $f(x) = 0$ for every x, which we assume not to be the case, the equation

$$f(x) = 0$$

has not more than n real roots; so if λ is an irrational root there exists a positive number δ such that the equation has no rational root between $\lambda - \delta$ and $\lambda + \delta$. Let p be an integer, and q a positive integer, and suppose first that $|(p/q) - \lambda| < \delta$. Then $f(p/q) \neq 0$, so that $|f(p/q)| \geqslant 1/q^n$; and if $x = p/q$ then $|x| < \delta + |\lambda|$, so that

$$|b_0 x^{n-1} + b_1 x^{n-2} + \ldots + b_{n-1}| \leqslant |b_0| (\delta + |\lambda|)^{n-1}$$
$$+ |b_1| (\delta + |\lambda|)^{n-2} + \ldots + |b_{n-1}|,$$

which is a positive number, c say, independent of x. Thus

$$1/q^n \leqslant |(p/q) - \lambda| c$$

provided that $|(p/q) - \lambda| < \delta$. If $|(p/q) - \lambda| \geqslant \delta$ then

$$|(p/q) - \lambda| \geqslant \delta/q^n.$$

Hence if k is the lesser of the positive numbers δ and $1/c$ then, in either case, $|(p/q) - \lambda| \geqslant k/q^n.$

(What we have proved here is a famous theorem due to the nineteenth-century French mathematician J. Liouville. It is the foundation of a substantial theory concerning irrational numbers that are 'badly approximable' by rational numbers. A real number is said to be *algebraic* if it is a root of a polynomial equation with integer coefficients, and to be *transcendental* otherwise; the algebraic numbers form a subfield of the field of real numbers, but one can deduce from Liouville's theorem, by an argument which he devised, that transcendental numbers exist in every open interval.)

30. By definition, $f^{-1}(E) = \{x : f(x) \in E\}$; this set is empty if the domain of f has no points x such that $f(x) \in E$. If $x \in f^{-1}(E \cup H)$ then $f(x) \in E \cup H$, that is $f(x)$ belongs to E or to H (or to both), so $x \in f^{-1}(E)$ or $x \in f^{-1}(H)$, which is to say that $x \in f^{-1}(E) \cup f^{-1}(H)$; thus

$$f^{-1}(E \cup H) \subseteq f^{-1}(E) \cup f^{-1}(H).$$

The opposite inclusion can be established similarly, by showing that if $x \in f^{-1}(E)$ or $x \in f^{-1}(H)$ then $x \in f^{-1}(E \cup H)$. In essentially the same way it can be shown that

$$f^{-1}(E \cap H) = f^{-1}(E) \cap f^{-1}(H).$$

Let X be the set of all numbers that are square roots of positive integers, and let f be the mapping $x \to x^2$ of X into the set Y of all positive integers. Then the range of f is the whole of Y, and if E is a subset of Y consisting of a single integer n then $f^{-1}(E)$ is the set consisting of the two numbers $\pm\sqrt{n}$ (so f has no inverse).

31. The domain of the function $f \circ \varphi$ is the domain of φ, namely the set Y; and for each y in Y, $(f \circ \varphi)(y)$ is defined to be $f(\varphi(y))$. Thus $f \circ \varphi$ maps Y into Y. The domain of $\varphi \circ f$ is the domain of f, namely X, and $\varphi \circ f$ maps X into X according to the formula

$$(\varphi \circ f)(x) = \varphi(f(x)) \quad (x \in X).$$

For any subset E of X, $(\varphi \circ f)(E)$ is a subset H of X, and $x \in H$ if and only if there is x' in E such that $(\varphi \circ f)(x') = x$. Thus if $E = \varnothing$ there is no x in H; that is to say,

$$(\varphi \circ f)(\varnothing) = \varnothing.$$

If E is a subset of Y then $y \in (f \circ \varphi)^{-1}(E)$ if and only if $f(\varphi(y)) \in E$. But $f(\varphi(y)) \in E$ if and only if $\varphi(y) \in f^{-1}(E)$; and $\varphi(y) \in f^{-1}(E)$ if and only if $y \in \varphi^{-1}(f^{-1}(E))$. Thus

$$(f \circ \varphi)^{-1}(E) = \varphi^{-1}(f^{-1}(E)).$$

(If $X \neq Y$ then $f \circ \varphi \neq \varphi \circ f$ because these functions have different domains. If $X = Y$ the functions may or may not be equal. For example, let X consist of the natural numbers, and let f be the mapping $x \to x^2$: if φ is the mapping $x \to x^3$ then $f \circ \varphi = \varphi \circ f$; but if φ is the mapping $x \to x+1$ then $f \circ \varphi \neq \varphi \circ f$. For any non-empty set X, we have $\circ$ as a binary operation for the set of all functions that map X into X; but, as is easily seen, this operation is not commutative unless X has only one element; it is, however, necessarily associative. The operation may be commutative within a particular set of functions: consider, for example, the set consisting of a function f, mapping X into X, together with all its *iterates*, $f \circ f$, $f \circ f \circ f$, …. . See **81.**)

32. The given equation holds when $s = t = 1$, so that

$$f(1) = f(1) + f(1),$$

and it follows immediately that $f(1) = 0$.

For any positive rational numbers s and t,

$$f(s) = f(st^{-1}t) = f(st^{-1}) + f(t),$$

so that $f(s) = f(t)$ if and only if $f(st^{-1}) = 0$; hence if this last equation can hold only when $st^{-1} = 1$, that is when $s = t$, then $f(s) \neq f(t)$ if $s \neq t$. Now

$$2f(s) = f(s) + f(s) = f(s^2),$$

so that if $2f(s) = f(2)$ then $f(s^2) = f(2)$; with the given assumption, this implies that $s^2 = 2$, and hence that s is not rational (see **10**).

(This last result can easily be generalized. If r is any positive rational number, it can be seen from the fundamental theorem of arithmetic that $r^{1/n}$ is irrational if n is a sufficiently large integer; then, if f satisfies the given conditions, there is no rational number s such that $f(s) = f(r)/n$. It follows that if P is the set of all positive rational numbers then $f(P)$ does not include P; for if $f(r) \in P$ then $f(r)/n \in P \setminus f(P)$. Moreover, $f(P)$ does not consist entirely of rational numbers: indeed, if s and t are distinct prime numbers then $f(s)$ or $f(t)$, or both, must be irrational; for if $f(s) = m/n$ and $f(t) = p/q$, where m, n, p, q are integers, then $f(s^{np}) = mp$ and $f(t^{mq}) = mp$, so that $s^{np} = t^{mq}$, which is impossible if s and t are distinct primes. Note that, under the given conditions, f establishes an isomorphism between the multiplicative group P and a subgroup of the additive group of all real numbers; we have shown that this subgroup neither includes nor is included in the subgroup consisting of the rational numbers. The existence of such a function f cannot, therefore, be proved within the context of elementary arithmetic. A proof of existence is supplied by the theory of the logarithmic function. See **43, 61**.)

33. (i) In any interval I with more than one point (any *non-degenerate* interval) we can choose rational numbers x_1, x_2 such that $x_1 < x_2$, and irrational numbers x_3, x_4 such that $x_3 < x_4$. Then, for the given function f,

$$f(x_1) - f(x_2) = x_1 - x_2 < 0,$$
$$f(x_3) - f(x_4) = -x_3 + x_4 > 0.$$

Thus f is not monotonic on I.

(ii) If x is rational then $f(f(x)) = f(x) = x$; if x is irrational then so is $-x$, and therefore $f(f(x)) = f(-x) = -(-x) = x$. Thus $f(f(x)) = x$ for every x, which is to say that f is inverse to itself.

(Note that if a monotonic function f is inverse to itself then either $f(x) = x$ for every x or $f(x) = -x$ for every x. To see this, suppose first that f is an increasing function: if x is such that $f(x) < x$ then

$$f(f(x)) < f(x),$$

which is impossible if $f(f(x)) = x$; if x is such that $f(x) > x$ then

$$f(f(x)) > f(x),$$

which is also impossible if $f(f(x)) = x$; thus $f(x) = x$ for every x if f is inverse to itself. If f is decreasing and inverse to itself a similar argument, with $-f$ in place of f, shows that $f(x) = -x$ for every x. Since the function f given in the exercise is neither of these two functions, we have an alternative proof that f is not monotonic.)

34. For the given function f, and any real number y, it is clear that $f(x) > |y|$, and hence $f(x) > y$, if $x > |y|^{\frac{1}{3}}$; and that $f(x) < -|y|$, and hence $f(x) < y$, if $x < -|y|^{\frac{1}{3}}$. Thus if I is the interval of all real numbers, $f(I)$ contains points on either side of y. But f is a continuous function, and therefore $f(I)$ is an interval; therefore $y \in f(I)$, which is to say that $y = f(x)$ for some x in I.

It is easy to see that f is an increasing function. Thus f has an inverse, f^{-1}; and f^{-1} is continuous because f is continuous. The domain of f^{-1} is the range of f, which consists of all real numbers. If f^{-1} could be represented by a polynomial, of degree n say, then $f \circ f^{-1}$ could be represented by a polynomial of degree $5n$. But $f \circ f^{-1}$ is represented by a polynomial of degree 1, and no function can be represented, on an interval with more than one point, by two distinct polynomials (because a polynomial equation can have at most finitely many roots, and an interval with more than one point has infinitely many points). Thus f^{-1} cannot be represented by a polynomial (on any non-degenerate interval).

35. Let
$$f(x) = x^3 + x - 1$$
for any real number x. Then f is an increasing function, and consequently there cannot be more than one value of x for which $f(x) = 0$. Now $f(x) = x(x^2 + 1) - 1$, so that

$$f(0 \cdot 68) = 0 \cdot 68 \times 1 \cdot 4624 - 1 = -0 \cdot 005568 < 0,$$

$$f(0 \cdot 69) = 0 \cdot 69 \times 1 \cdot 4761 - 1 = 0 \cdot 018509 > 0.$$

Since f is continuous, it follows that $f(x) = 0$ for some x between $0 \cdot 68$ and $0 \cdot 69$.

(We have calculated $f(0 \cdot 68)$ and $f(0 \cdot 69)$ with unnecessary accuracy. It would have been enough to observe that

$$f(0 \cdot 68) < 0 \cdot 68 \times 1 \cdot 47 - 1 = -0 \cdot 0004 < 0,$$

$$f(0 \cdot 69) > 0 \cdot 69 \times 1 \cdot 47 - 1 = 0 \cdot 0143 > 0.$$

Computational labour can often be saved in this sort of way. For an 'algebraic' solution of the given equation, see **85**.)

36. We must assume that y is fully specified numerically, as a rational number or as a number that can be rationally calculated to any degree of accuracy. If x is a rational number in I it may happen that $f(x) = y$, but if $f(x) \neq y$ we can compute $f(x)$ with sufficient accuracy to decide whether $f(x) < y$ or $f(x) > y$. Suppose that, after a few trials (chosen, perhaps, by shrewd judgment), we find rational numbers ξ_1 and ξ_2 in I such that

$$f(\xi_1) < y < f(\xi_2).$$

Since f is an increasing function, $\xi_1 < \xi_2$; and since f is continuous, the open interval $]\xi_1, \xi_2[$ has a point x_0 such that $f(x_0) = y$. There is only one such point x_0, and if

$$\xi_3 = \xi_1 + \tfrac{1}{3}(\xi_2 - \xi_1) \quad \text{and} \quad \xi_4 = \xi_1 + \tfrac{2}{3}(\xi_2 - \xi_1)$$

we can decide whether x_0 lies in $]\xi_1, \xi_4[$ or in $]\xi_3, \xi_2[$ (it may, of course, lie in both these intervals). Thus we locate x_0 within an interval of length $\tfrac{2}{3}(\xi_2 - \xi_1)$. Repeating this procedure, we can eventually show that x_0 lies in an open interval, say $]x_1, x_2[$, of length $(\tfrac{2}{3})^n(\xi_2 - \xi_1)$, for any given positive integer n. But if n is sufficiently large then $(\tfrac{2}{3})^n < \epsilon/(\xi_2 - \xi_1)$, because

$$\inf\{(\tfrac{2}{3})^n \colon n = 1, 2, 3, \ldots\} = 0$$

(see **17**). Then $|x_1 - x_2| < \epsilon$, and, since $x_1 < x_0 < x_2$,

$$f(x_1) < y < f(x_2).$$

Since $f^{-1}(y) = x_0$, this process locates $f^{-1}(y)$ within the interval $]x_1, x_2[$; so that any rational number in this interval represents $f^{-1}(y)$ with an error less than the given positive number ϵ, and the mid-point of the interval gives an error less than $\tfrac{1}{2}\epsilon$.

(This is not necessarily the best method of computing $f^{-1}(y)$ in a particular case. More-refined methods are available for functions that satisfy special conditions. Any method is legitimate provided that it leads to numbers x_1 and x_2 that are shown to satisfy the required conditions.)

37. The given information implies that

$$f(3\cdot16) < 4\cdot9975 \quad \text{and} \quad f(3\cdot17) \geqslant 5\cdot0105.$$

Since f is continuous, it follows that there is a number x between $3\cdot16$ and $3\cdot17$ such that $f(x) = 5$. Since f is increasing, there is only one such x, and this is $f^{-1}(5)$. Thus

$$3\cdot16 < f^{-1}(5) < 3\cdot17.$$

If $f(3\cdot16)$ and $f(3\cdot17)$ had been specified to 2 decimal places only, as $5\cdot00$ and $5\cdot01$ respectively, we should have known only that

$$4\cdot995 \leqslant f(3\cdot16) < 5\cdot005 \quad \text{and} \quad 5\cdot005 \leqslant f(3\cdot17) < 5\cdot015.$$

From these inequalities we cannot infer that there is a number x such that $f(x) = 5$; if this fact is known in some other way, we can infer that $f^{-1}(5) < 3\cdot17$ but not that $f^{-1}(5) > 3\cdot16$.

38. Suppose that $0 \leqslant x_1 \leqslant x_2 \leqslant 1$. Then $[0, x_1] \subseteq [0, x_2]$, and therefore $f([0, x_1]) \subseteq f([0, x_2])$. Since, for any x in $[0, 1]$,

$$\varphi(x) = \sup f([0, x]),$$

it follows that $\varphi(x_1) \leqslant \varphi(x_2)$. Thus φ is a non-decreasing function.

Now suppose that f is continuous, and let ϵ be any positive number. Then there is a positive number δ such that if $x \in [0, 1]$ and $|x - x_1| < \delta$ then $|f(x) - f(x_1)| < \epsilon$. Hence if $|x_1 - x_2| < \delta$ and $0 \leqslant x \leqslant x_2$ then either $x \leqslant x_1$, in which case $f(x) \leqslant \varphi(x_1)$ by the definition of $\varphi(x_1)$, or $|x - x_1| < \delta$, in which case $f(x) < f(x_1) + \epsilon$; in each case $f(x) < \varphi(x_1) + \epsilon$; therefore $\varphi(x_2) \leqslant \varphi(x_1) + \epsilon$ by the definition of $\varphi(x_2)$. Since $\varphi(x_1) \leqslant \varphi(x_2)$, it follows that if $|x_1 - x_2| < \delta$ then $|\varphi(x_1) - \varphi(x_2)| < \epsilon$. Thus φ is continuous.

(It is clear that the domain of f could be an interval other than $[0, 1]$. If f is continuous then $\varphi(x)$ is the maximum value attained by f on the interval $[0, x]$. Instruments that register $f(x)$ as a physical magnitude at time x—thermometers for example—are sometimes arranged to register $\varphi(x)$ simultaneously. If 'sup' is replaced by 'inf' in the definition of φ we obtain a non-increasing function, continuous if f is continuous. The continuity of φ does not imply that of f; see **83**.)

39. It is clear that $f(0) = 0$, that $f(x) > 0$ if $x \neq 0$, and that, for every real number x,
$$f(-x) = f(x).$$

Thus the curve represented by the equation

$$y = f(x)$$

lies above the x-axis, except that it contains the point $(0, 0)$, and it is symmetric about the y-axis. Its shape can be indicated by plotting, to a suitable scale, the points (x, y) for which

$$y = \tfrac{1}{5}n, \quad n = 0, 1, \ldots, 10.$$

If $f_1(x) = |x|$ for every real x, and if $f_2(x) = x^{\frac{1}{4}}$ for every non-negative x, then $f = f_2 \circ f_1$. Now the function f_1 is continuous by virtue of the triangle inequality, and f_2 is continuous because it is the inverse of a continuous univalent function. Since composites of continuous functions are continuous, it follows that f is continuous.

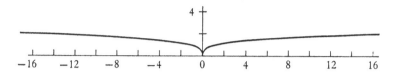

40. If $x = -1$ then $x^2 = 1$ so $[x^2] = 1$.

If $-1 < x < 1$ then $0 \leqslant x^2 < 1$ so $[x^2] = 0$.

If $1 \leqslant x < \sqrt{2}$ then $1 \leqslant x^2 < 2$ so $[x^2] = 1$.

If $\sqrt{2} \leqslant x < \sqrt{3}$ then $2 \leqslant x^2 < 3$ so $[x^2] = 2$.

If $\sqrt{3} \leqslant x < 2$ then $3 \leqslant x^2 < 4$ so $[x^2] = 3$.

If $x = 2$ then $x^2 = 4$ so $[x^2] = 4$.

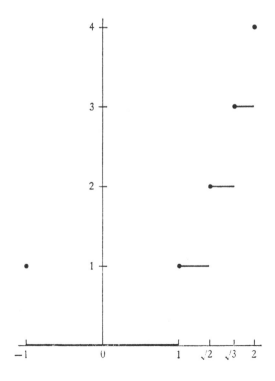

The graph thus consists of the points $(-1, 1)$ and $(2, 4)$, the points of the x-axis between $(-1, 0)$ and $(1, 0)$, and all but the right-hand end-points of the three straight segments joining respectively $(1, 1)$ and $(\sqrt{2}, 1)$, $(\sqrt{2}, 2)$ and $(\sqrt{3}, 2)$, $(\sqrt{3}, 3)$ and $(2, 3)$. The left-hand end-points of these three segments are marked by dots on the diagram to show that they belong to the graph. (These three dots, and the other two, represent points on the graph of the function $x \to x^2$.)

On each of the intervals $]-1, 1[$, $[1, \sqrt{2}[$, $[\sqrt{2}, \sqrt{3}[$, $[\sqrt{3}, 2[$, the function $x \to [x^2]$ has a constant value; it is therefore integrable over each of the corresponding closed intervals, and each integral is the product of the constant value of the function and the length of the interval. By the additivity of the integration process, the function is therefore integrable over the interval $[-1, 2]$, and

$$\int_{-1}^{2} [x^2]\, dx = 0 + (\sqrt{2} - 1) + 2(\sqrt{3} - \sqrt{2}) + 3(2 - \sqrt{3})$$
$$= 5 - \sqrt{2} - \sqrt{3}.$$

41. For $n = 1, 2, 3, \ldots$, let $P(n)$ be the proposition

$$1^2 + 2^2 + \ldots + n^2 = \tfrac{1}{6}n(n+1)(2n+1).$$

Evidently $P(1)$ is true. If $P(n)$ is true then

$$1^2 + 2^2 + \ldots + (n+1)^2 = \tfrac{1}{6}n(n+1)(2n+1) + (n+1)^2$$
$$= \tfrac{1}{6}(n+1)(n+2)(2n+3);$$

thus $P(n)$ implies $P(n+1)$. Accordingly, by the principle of mathematical induction, $P(n)$ is true for every n.

Let the interval $[0, 1]$ be divided into n closed subintervals by the points r/n, where $r = 0, 1, 2, \ldots, n$. If $(r-1)/n \leqslant x \leqslant r/n$, where $r \geqslant 1$, then x^2 has minimum value $(r-1)^2/n^2$ and maximum value r^2/n^2. Accordingly, if

$$S_n = (1^2 + 2^2 + \ldots + n^2)/n^3 \quad \text{and} \quad s_n = (0^2 + 1^2 + \ldots + (n-1)^2)/n^3$$

then S_n and s_n are respectively upper and lower Riemann sums for $\int_0^1 x^2\, dx$. But, by the propositions $P(n)$ and $P(n-1)$,

$$S_n = \frac{1}{3} + \frac{1}{2n}\left(1 + \frac{1}{3n}\right) \quad \text{and} \quad s_n = \frac{1}{3} - \frac{1}{2n}\left(1 - \frac{1}{3n}\right).$$

Hence $s_n \leqslant \tfrac{1}{3} \leqslant S_n$ and $S_n - s_n = 1/n$. It follows that $\int_0^1 x^2\, dx = \tfrac{1}{3}$.

$\Big($This result can be generalized in the following way. If k is any positive integer and c is any real number, and if

$$S_n = (1^k + 2^k + \ldots + n^k)(c/n)^{k+1}$$

and $$s_n = (0^k + 1^k + \ldots + (n-1)^k)(c/n)^{k+1},$$

then S_n and s_n are respectively upper and lower Riemann sums for $\int_0^c x^k \, dx$; and $S_n - s_n = c^{k+1}/n$, which we can make as small as we like by taking n sufficiently large (see **67**). Now the well-known identity

$$a^{k+1} - b^{k+1} = (a-b)(a^k + a^{k-1}b + \ldots + b^k)$$

shows that, for any number r,

$$r^{k+1} - (r-1)^{k+1} = r^k + r^{k-1}(r-1) + \ldots + (r-1)^k,$$

where there are $k+1$ terms on the right. If $r \geqslant 1$ none of these $k+1$ terms is less than $(r-1)^k$ or greater than r^k; therefore

$$(r-1)^k \leqslant \frac{r^{k+1} - (r-1)^{k+1}}{k+1} \leqslant r^k$$

if $k = 1, 2, \ldots, n$, and consequently

$$s_n \leqslant \left(\frac{1^{k+1} - 0^{k+1}}{k+1} + \ldots + \frac{n^{k-1} - (n-1)^{k+1}}{k+1} \right) \left(\frac{c}{n} \right)^{k+1} \leqslant S_n,$$

that is, $s_n \leqslant \dfrac{c^{k+1}}{k+1} \leqslant S_n$. It now follows that

$$\int_0^c x^k \, dx = \frac{c^{k+1}}{k+1}.$$

From this result, and the linearity and additivity of the integration process, we can integrate any polynomial function over any bounded interval. In particular, it is easy to show that for any real numbers a, b, c, and any non-negative integer k,

$$\int_b^c (x-a)^k \, dx = \frac{(c-a)^{k+1} - (b-a)^{k+1}}{k+1}.\Big)$$

42. If $0 \leqslant x \leqslant 1$,

$$\{(1 + \tfrac{1}{2}x)\sqrt{(1-x)}\}^2 = (1 + x + \tfrac{1}{4}x^2)(1-x) = 1 - \tfrac{3}{4}x^2 - \tfrac{1}{4}x^3 \leqslant 1,$$

$$\{(1 + \tfrac{2}{3}x)\sqrt{(1-x)}\}^2 = (1 + \tfrac{4}{3} + \tfrac{4}{9}x^2)(1-x) = 1 + \tfrac{1}{9}x(3 - 8x - 4x^2).$$

Now if $0 \leqslant x \leqslant \tfrac{1}{4}$ then $8x + 4x^2 < 3$; therefore, for these values of x,

$$(1 + \tfrac{1}{2}x)\sqrt{(1-x)} \leqslant 1 \leqslant (1 + \tfrac{2}{3}x)\sqrt{(1-x)},$$

whence the required inequalities follow. If $0 \leqslant x \leqslant \frac{1}{2}$ then $0 \leqslant x^2 \leqslant \frac{1}{4}$, and therefore

$$1 + \tfrac{1}{2}x^2 \leqslant \frac{1}{\sqrt{(1-x^2)}} \leqslant 1 + \tfrac{2}{3}x^2.$$

Hence, by the linearity and positivity of the integration process,

$$\int_0^{\frac{1}{2}} (1 + \tfrac{1}{2}x^2) \, dx \leqslant \int_0^{\frac{1}{2}} \frac{dx}{\sqrt{(1-x^2)}} \leqslant \int_0^{\frac{1}{2}} (1 + \tfrac{2}{3}x^2) \, dx,$$

that is, since $\int_0^{\frac{1}{2}} x^2 \, dx = \frac{1}{24}$,

$$\frac{25}{48} \leqslant \int_0^{\frac{1}{2}} \frac{dx}{\sqrt{(1-x^2)}} \leqslant \frac{19}{36}.$$

43. By definition, if $x > 0$,

$$\log x = \int_1^x \frac{du}{u}.$$

Hence if $t \geqslant 0$,

$$\log(1+t) = \int_1^{1+t} \frac{du}{u} = \int_1^{1+t} \frac{du}{1+(u-1)}.$$

Now if n is any positive integer, and $a \neq -1$,

$$\frac{1}{1+a} = 1 - a + a^2 - \dots + a^{2n} - \frac{a^{2n+1}}{1+a}.$$

Using this well-known identity, with $a = u - 1$, and using the fact that

$$\int_1^{1+t} (u-1)^k \, du = \frac{t^{k+1}}{k+1}$$

for any positive integer k, we find from the linearity of the integration process that

$$\log(1+t) = t - \tfrac{1}{2}t^2 + \tfrac{1}{3}t^3 - \dots + \frac{1}{2n+1}t^{2n+1} - \int_1^{1+t} \frac{(u-1)^{2n+1} \, du}{u}.$$

But, by the positivity of the integration process,

$$0 \leqslant \int_1^{1+t} \frac{(u-1)^{2n+1} \, du}{u} \leqslant \int_1^{1+t} (u-1)^{2n+1} \, du = \frac{t^{2n+2}}{2n+2}.$$

It follows that the number

$$t - \tfrac{1}{2}t^2 + \tfrac{1}{3}t^3 - \dots + \frac{1}{2n+1}t^{2n+1}$$

exceeds $\log(1+t)$ by a number which is not greater than $t^{2n+2}/(2n+2)$. (In fact, if $t > 0$ the excess is positive and less than $t^{2n+2}/(2n+2)$, by the strict positivity of the integration process for continuous functions.)

Using this result with $t = \frac{1}{4}$ and $n = 1$, we find that $\frac{1}{4} - \frac{1}{32} + \frac{1}{192}$, which is $\frac{43}{192}$, exceeds $\log \frac{5}{4}$ by not more than $\frac{1}{1024}$, and so by less than 10^{-3}.

(More generally, this method enables us to calculate $\log(1 + (1/k))$, which is $\log(k + 1) - \log k$, to any degree of accuracy, for any positive integer k. Since $\log 1 = 0$, we can thus calculate $\log 2$, $\log 3$, and so on in succession, with an error which can be kept below a prescribed bound. However, to calculate $\log 2$ with high accuracy by this method it is necessary for n to be very large, and for this initial step it is better to use a more sophisticated method, or even the crude method of computing Riemann sums.)

44. Let R be the set of all real numbers.

(i) We know that the mapping $x \to x$, of R into R, is continuous. (This, of course, is an immediate consequence of the definition of continuity.) Products of continuous functions are continuous: hence, for each natural number n, the mapping $x \to x^n$ is continuous. Also, sums of continuous functions are continuous: hence if $f(x) = x^4 + x^3 + x^2$ for every x in R then f is continuous. If $x > 1$ then $f(x) > x$, so f is unbounded (though, since $f(x) = x^2\{(x + \frac{1}{2})^2 + \frac{3}{4}\}$, f is bounded below).

(ii) Since every constant function is continuous, we can show as in case (i) that every polynomial function is continuous; in particular, the mapping $x \to 1 + x^2$, of R into R, is continuous. If $x \in R$ then $1 + x^2 \neq 0$: hence, by a theorem on quotients of continuous functions, if

$$f(x) = x/(1 + x^2)$$

then f is continuous. If $x \in R$ then $2|x| \leqslant 1 + x^2$ (equality occurring when $x = \pm 1$): therefore $-\frac{1}{2} \leqslant f(x) \leqslant \frac{1}{2}$, so that f is bounded.

(iii) The logarithmic function is continuous, and its domain includes the range of the mapping $x \to 1 + x^2$. Since composites of continuous functions are continuous, the mapping $x \to \log(1 + x^2)$, of R into R, is continuous. Hence if $f(x) = x \log(1 + x^2)$ then f, being a product of continuous functions, is continuous. Since $\log(1 + x^2) \geqslant \log 2$ if $x > 1$, f is unbounded.

(iv) The exponential function is continuous (being the inverse of a continuous univalent function): hence if $f(x) = e^{-x^2}$ for every x in R then f, being a composite of continuous functions, is continuous. The range of the exponential function (of a real variable) is the set of all positive numbers: therefore, if $x \in R$, $f(x) > 0$. The exponential function is an increasing function, and its value at 0 is 1: therefore, since $-x^2 < 0$, $f(x) \leqslant 1$. Thus f is bounded.

(We know that the image of an interval by a continuous function is necessarily an interval: hence the range, $f(R)$, of each of the functions f considered here is an interval. In case (i), $f(R) = \{y: y \geqslant 0\}$; in case (ii), $f(R) = [-\frac{1}{2}, \frac{1}{2}]$; in case (iii), $f(R) = R$; in case (iv), $f(R) =]0, 1]$.)

45. We know that the logarithmic function is univalent, and that its range is the set of all real numbers. Its domain is the set of all positive numbers. Thus we know that it has an inverse function, and that this maps the set of all real numbers on to the set of all positive numbers. This inverse function is, by definition, *the exponential function* (of a real variable), whose value at x is denoted by e^x. Since, if $x > 0$,

$$\log x = \int_1^x \frac{\mathrm{d}u}{u},$$

a fundamental theorem of the differential calculus tells us that

$$\frac{\mathrm{d}}{\mathrm{d}x} \log x = \frac{1}{x}.$$

Another such theorem (on the derivative of an inverse function) now tells us that, if x is any real number,

$$\frac{\mathrm{d}}{\mathrm{d}x} e^x = 1/(1/e^x) = e^x.$$

Let a, b, c, α, β, γ be real numbers such that, for every number x in a given open interval I,

$$\alpha e^{ax} + \beta e^{bx} + \gamma e^{cx} = 0. \tag{1}$$

Then, for every x in I,

$$\frac{\mathrm{d}}{\mathrm{d}x} (\alpha e^{ax} + \beta e^{bx} + \gamma e^{cx}) = 0,$$

that is, by fundamental rules of the differential calculus,

$$a\alpha e^{ax} + b\beta e^{bx} + c\gamma e^{cx} = 0. \tag{2}$$

Consequently, for the same reasons,

$$a^2 \alpha e^{ax} + b^2 \beta e^{bx} + c^2 \gamma e^{cx} = 0 \tag{3}$$

for every x in I. By a standard process of elimination, or by the theory of 3-by-3 determinants, we deduce from equations (1), (2), and (3) that either $(a-b)(b-c)(c-a) = 0$ or, for every x in I, $\alpha e^{ax} = \beta e^{bx} = \gamma e^{cx} = 0$. The first of these alternatives is excluded if a, b, c are distinct; and since 0 is not a value of the exponential function, the second alternative implies that $\alpha = \beta = \gamma = 0$.

(We have proved that a certain set of 3 functions is 'linearly independent'. There is a similar result for n functions of the same type.)

46. We know that since f is differentiable at every point of $[a, b]$ it is continuous on $[a, b]$. Hence we know that $f([a, b])$ is a bounded closed interval. The right-hand end-point of $f([a, b])$ is $f(x_0)$, where x_0 is a point (not necessarily unique) of $[a, b]$. Thus $f(x) \leqslant f(x_0)$ for every x in $[a, b]$. Hence if $x_0 > a$ and $a \leqslant x < x_0$ then

$$\frac{f(x) - f(x_0)}{x - x_0} \geqslant 0,$$

whence it follows, by the definition of $f'(x_0)$, that $f'(x_0) \geqslant 0$. If $x_0 < b$ and $x_0 < x \leqslant b$ then

$$\frac{f(x) - f(x_0)}{x - x_0} \leqslant 0,$$

and therefore $f'(x_0) \leqslant 0$.

(This discussion shows that if the differentiable function f on $[a, b]$ takes its greatest value at a point x_0 which is neither a nor b then $f'(x_0) = 0$. It is easy to see, from a simple example, that f may take its greatest value at b although $f'(b) \neq 0$. Similar considerations apply in the case of the least value. And of course f' may take the value 0 at a point x of $[a, b]$ such that $f(x)$ is neither the greatest nor the least of the values taken by f on $[a, b]$.)

47. If φ has a local minimum at a point x of I which is not an end-point then, by the definition of the term 'local minimum', I has an open subinterval I' which contains x and is such that $\varphi(x') \geqslant \varphi(x)$ for every point x' of I'. In this case, therefore, the incremental ratio

$$\frac{\varphi(x') - \varphi(x)}{x' - x}$$

is non-negative if $x' > x$ and is non-positive if $x' < x$, provided that $x' \in I'$. But if φ is differentiable at x, and ϵ is any positive number, then by the definition of $\varphi'(x)$,

$$\varphi'(x) - \epsilon < \frac{\varphi(x') - \varphi(x)}{x' - x} < \varphi'(x) + \epsilon$$

provided only that $|x' - x|$ is sufficiently small and not 0. It follows that $\varphi'(x)$ cannot be greater than ϵ or less than $-\epsilon$; so, since ϵ is an arbitrary positive number, the only possibility is that $\varphi'(x) = 0$.

Now let φ be defined on $[a, b]$ by the equation

$$\varphi(x) = f(x) - g(x),$$

where f and g satisfy the given conditions. Then $\varphi(a) = \varphi(b) = 0$, and φ is continuous on $[a, b]$ and differentiable at each point of $]a, b[$. Because of its continuity on $[a, b]$, φ has a least value, c say; and φ has a local minimum at any point x of $[a, b]$ for which $\varphi(x) = c$. If there is such a point x in $]a, b[$ then $\varphi'(x) = 0$ and therefore $f'(x) = g'(x)$. If not, then φ takes the value c only at a and b, so that $c = 0$ and $\varphi(x) \geqslant 0$ for every x in $]a, b[$; in this case there must evidently be a point x of $]a, b[$ at which the function $-\varphi$ has a local minimum, and for this point we conclude as before that $f'(x) = g'(x)$.

(Note that this argument gives a proof of the mean-value theorem if we take g to be defined from f by the equation

$$g(x) = f(a) + (x-a)\frac{f(b)-f(a)}{b-a}.$$

Note too that the functions considered need not be differentiable at the end-points: for example, f satisfies the stated conditions if

$$f(x) = (x-a)^{\frac{1}{2}}(b-x)^{\frac{1}{2}} \quad (a \leqslant x \leqslant b),$$

but in this case f is not differentiable at a or at b.)

48. Since I is an interval and a and b are points of I,

$$[a, b] \subseteq I.$$

Therefore f, being differentiable throughout I, is differentiable throughout $[a, b]$. Hence if $a \leqslant x \leqslant b$ then g is differentiable at x, and

$$g'(x) = f'(x) - \eta.$$

Since η lies between $f'(a)$ and $f'(b)$ it follows that 0 lies between the numbers $g'(a)$ and $g'(b)$, so that one of these numbers is positive and the other negative.

Now since g is differentiable throughout $[a, b]$ it is continuous on $[a, b]$; therefore the set $g([a, b])$ has a least element and a greatest element. Let x_0 and x_0' be points of $[a, b]$ such that $g(x_0) = \inf g([a, b])$ and $g(x_0') = \sup g([a, b])$. If $x_0 > a$, in particular if $x_0 = b$, then $g'(x_0) \leqslant 0$; and if $x_0 = a$ then $g'(x_0) \geqslant 0$ (see **46**). Similarly, if $x_0' = a$ then $g'(x_0') \leqslant 0$ and if $x_0' = b$ then $g'(x_0') \geqslant 0$. Thus, since $g'(a)$ and $g'(b)$ are not both positive or both negative, x_0 and x_0' cannot both be end-points of $[a, b]$: one of them, which we may call ξ, must be in the open interval $]a, b[$. Since $g(\xi)$ is either $\inf g([a, b])$ or $\sup g([a, b])$, g has either a local minimum or a local maximum at ξ; therefore, since ξ is not an end-point of $[a, b]$, $g'(\xi) = 0$ (see **47**). Thus $f'(\xi) = \eta$. Since η is an arbitrary number between $f'(a)$ and $f'(b)$, it follows that the set $f'(I)$ is an interval.

(Note that the function f' need not be continuous: for example, if $f(x) = x^2 \sin(1/x)$ for every real number x other than 0, and if $f(0) = 0$, then f is differentiable everywhere but f' is not continuous on any non-degenerate interval containing 0. A function that maps intervals to intervals is said to have *the Darboux property*, after the nineteenth-century French mathematician G. Darboux; this property represents an intuitive idea of continuity, and that it does not imply continuity is one of the surprising facts of analysis. See **86**.)

49. *The mean-value theorem* asserts that, under the stated conditions, there is a number ξ in the open interval $]a, b[$ such that

$$\frac{f(b) - f(a)}{b - a} = f'(\xi).$$

The conditions (i) and (ii) enable us to apply the mean-value theorem to the case in which f is the function sin, $a = 0$, and $b = \theta$, where θ can be any number between 0 and $\frac{1}{2}\pi$. Since $\sin 0 = 0$, and the derivative of sin is cos, we see that there is a number ξ (depending on the choice of θ) such that $0 < \xi < \theta$ and
$$\frac{\sin \theta}{\theta} = \cos \xi.$$

Since $\sin \theta$ increases from 0 to 1 as θ increases from 0 to $\frac{1}{2}\pi$, condition (iii) shows that $\cos^2 \theta$ decreases from 1 to 0, and condition (ii) shows that $\cos \theta$ is non-negative for these values of θ; therefore $\cos \theta$ decreases from 1 to 0 as θ increases from 0 to $\frac{1}{2}\pi$. Accordingly, if $0 < \xi < \theta$ then $\cos \theta < \cos \xi < 1$. Thus, for any number θ between 0 and $\frac{1}{2}\pi$,

$$\cos \theta < \frac{\sin \theta}{\theta} < 1.$$

(A special case of the mean-value theorem, in which it is assumed that $f(a) = f(b) = 0$, is known as 'Rolle's theorem', after Michel Rolle, a French mathematician of the late seventeenth century, who was one of the first critics of the unsound arguments used by early practioners of the calculus; for his theorem he considered only polynomial functions, so that it was, in effect, a theorem about the real roots of polynomial equations with real coefficients. The importance of the mean-value theorem was not fully understood before the nineteenth century. A simple example shows that the theorem can be false for complex-valued functions—a fact which is easily overlooked.)

50. The polynomials $(x+3)^2$ and $(x+2)^3$ have no common factor. Hence there are polynomials $p(x)$ and $q(x)$ which satisfy the identity

$$x^3 + 4x^2 - 10 = p(x)(x+3)^2 + q(x)(x+2)^3;$$

and since the degree of the polynomial $x^3 + 4x^2 - 10$ is less than that of $(x+3)^2(x+2)^3$, we can assume that the degrees of $p(x)$ and $q(x)$ are at most 2 and at most 1 respectively. These polynomials are then uniquely determined, and they can be calculated in the following way. Let

$$p(x) = a_0(x+2)^2 + a_1(x+2) + a_2 \quad \text{and} \quad q(x) = b_0(x+3) + b_1.$$

Since $a_2 = p(-2)$, a simple computation shows that $a_2 = -2$; and since $b_1 = q(-3)$ we similarly find that $b_1 = 1$. Hence

$$\{a_0(x+2)^2 + a_1(x+2)\}(x+3)^2 + b_0(x+3)(x+2)^3 = 0$$

and therefore
$$\{a_0(x+2) + a_1\}(x+3) + b_0(x+2)^2 = 0.$$

From this identity we find, on taking x to be -2, then -3, then any other number, that $a_1 = 0$, that $b_0 = 0$, and finally that $a_0 = 0$. It follows that if x is not -2 or -3 then

$$\frac{x^3 + 4x^2 - 10}{(x+3)^2(x+2)^3} = \frac{1}{(x+3)^2} - \frac{2}{(x+2)^3}.$$

The interval $[-1, 1]$ does not contain -2 or -3, and therefore

$$\int_{-1}^{1} \frac{x^3 + 4x^2 - 10}{(x+3)^2(x+2)^3}\,dx = \int_{-1}^{1} \frac{dx}{(x+3)^2} - 2\int_{-1}^{1} \frac{dx}{(x+2)^3}$$

$$= \tfrac{1}{4} - \tfrac{8}{9} = -\tfrac{23}{36}.$$

(We have illustrated a systematic method of 'resolution into partial fractions'. Other methods may be used in particular cases, the only criterion of legitimacy being the correctness of the result. To guard against errors, of reasoning or of computation, one should always check the correctness of the result by 'multiplying out' and then comparing coefficients.)

HINTS AND COMMENTS FOR
EXERCISES 51–100

51. $B \setminus A$ has 12 elements $(28-16)$.

$C \setminus (A \cup B)$ has 9 elements $(27-(14+13-9))$.

Hence $A \cup B \cup C$ has 64 elements $(43+12+9)$.

(This is an exercise on the additivity of number: if E and H are disjoint sets having m elements and n elements respectively then $E \cup H$ has $m+n$ elements. In a more general form, this notion of additivity is the basis of the important branch of mathematics known as 'measure theory', which includes the foundations of the theory of probability.)

52. Two finite sets can be put into one-to-one correspondence with each other if and only if they have the same number of elements; and if A has m elements and B has n elements, then $A \times B$ has mn elements. Thus the three results correspond respectively to the commutative law, the associative law, and the cancellation law, for multiplication of natural numbers.

53. The 'if' implication is easy to establish. For the 'only if' implication, suppose that $\sim$, as defined in terms of Q, is an equivalence relation. Then $\sim$ is reflexive, and from this it easily follows that $1 \in Q$. Suppose that $q \in Q$ and $q' \in Q$, where $q < q'$. Then $q'q \sim q$, $q \sim 1$, and $1 \sim q'$; hence, since $\sim$ is transitive, $q'q \sim 1$ and $q \sim q'$, and therefore $q'q \in Q$ and $q'/q \in Q$. It follows that if Q contains numbers other than 1, and if h is the least of these, then Q contains $1, h, h^2, \ldots$ and no other numbers.

54. Use the sieve of Eratosthenes.

55. If the sum were an integer, s say, then $p_1 p_2 \ldots p_n s$ would be an integer divisible by p_1. But

$$p_1 p_2 \ldots p_n s = p_2 p_3 \ldots p_n + p_1 r,$$

where r is an integer; and $p_2 p_3 \ldots p_n$ is not divisible by p_1.

56. *Binary operation*, Ω, on a non-empty set E: mapping, $(x, y) \rightarrow x \Omega y$, of $E \times E$ into E.

Commutative law, for a binary operation Ω on a set E:

$$x\,\Omega\,y = y\,\Omega\,x$$

for any elements x, y of E.

Associative law, for a binary operation Ω on E:

$$(x\,\Omega\,y)\,\Omega\,z = x\,\Omega\,(y\,\Omega\,z) \quad (= x\,\Omega\,y\,\Omega\,z)$$

for any elements x, y, z of E.

(i) For (say) the set of non-negative integers, the binary operation

$$(x, y) \to |x - y|$$

is commutative but not associative.

(ii) For any set having more than one element, the binary operation

$$(x, y) \to x$$

is associative but not commutative. (See also **27** and **77**.)

(A binary operation that obeys the commutative law is an example of a symmetric function of two variables; scalar multiplication, in vector algebra, is another example (see **23**). Vector multiplication (see **24**) is a binary operation which is neither commutative nor associative.)

57. *Group:* set G with an associative operation Ω (making a semigroup —see **103**) such that for any elements a, b of G there are elements x, y of G satisfying the equations

$$a\,\Omega\,x = b = y\,\Omega\,a.$$

Abelian group: group whose operation obeys the commutative law.

(A group—or a semigroup—may be written in *the multiplicative notation,* in which xy is written for $x\,\Omega\,y$, or, if the operation is commutative, in *the additive notation,* in which $x + y$ is written for $x\,\Omega\,y$. Every group has a unique element u such that

$$u\,\Omega\,x = x = x\,\Omega\,u$$

for every element x of the group; in the multiplicative notation this 'neutral' element u is often denoted by 1 and referred to as *the unit element,* and in the additive notation it is usually denoted by 0 and referred to as *the zero element* or simply 'zero'. Every element x of a group has a unique *inverse,* which is an element y of the group such that

$$x\,\Omega\,y = u = y\,\Omega\,x;$$

this inverse is denoted by x^{-1} in the multiplicative notation, by $-x$ in additive notation. The name 'Abelian group' commemorates the nineteenth-century Norwegian mathematician N. H. Abel.)

Ring: set R with two associative operations, called addition and multiplication (and denoted accordingly) such that R with addition is an Abelian group (the *additive group* of the ring) and

$$x(y+z) = xy+xz \quad \text{and} \quad (x+y)z = xz+yz$$

for any elements x, y, z of R. (The requirements expressed by these two identities are respectively *the left-hand distributive law* and *the right-hand distributive law*, which coalesce into *the distributive law* in the case of a *commutative ring*—a ring in which multiplication obeys the commutative law. See **81**.)

(Every ring has a unique *zero element*, namely that of its additive group. It may or may not have a *unit element*—multiplicatively neutral—and it cannot have more than one such element (see **103**). Any ring can be regarded as a subring of a ring which has a unit element (see **108**).)

The element p is the zero element of the ring (see **9**). Any Abelian group can be made into a ring by writing it in the additive notation and defining multiplication by the rule that all products are to be zero. Less trivially, consider the multiples of 4 in the ring of even integers modulo 8.

58. The ring $R_1 \times R_2$ has divisors of zero: if $x_1 \neq 0_1$ in R_1 and $x_2 \neq 0_2$ in R_2 (where 0_1 and 0_2 are the zero elements of R_1 and of R_2 respectively), then $(0_1, x_2)$ and $(x_1, 0_2)$ are non-zero elements of $R_1 \times R_2$, and their product is zero. Such a pair of elements cannot exist in a field. For an element (x_1, x_2) of $R_1 \times R_2$ to have a reciprocal in $R_1 \times R_2$ it is necessary and sufficient that x_1 and x_2 have reciprocals in R_1 and in R_2 respectively.

59. $R_2 = \{(1-u)x: x \in R\}$.

(An element u such that $u^2 = u$ is said to be *idempotent*. In a ring that has a unit element, u is idempotent if and only if $1-u$ is idempotent; and 1 is the only idempotent element having a reciprocal. See **106**.)

60. In the field of integers modulo 2, each element has exactly one square root. If each of n elements of a field has two square roots in the field then the field has at least $2n+1$ elements; hence it is impossible for every non-zero element of a finite field to have two square roots in the field.

61. If $a+a = 0$ and $a \neq 0$ then $1+1 = 0$, since $a+a = a(1+1)$; hence $x+x = 0$.

Let $x \to x'$ be an isomorphism of G into the additive group of F. Then $1' = 0$. If $(-1)' = 0$ then $-1 = 1$ since the mapping is one-to-one; if

$(-1)' = a \neq 0$ then $a \dotplus a = 0$ since $(-1)^2 = 1$. Hence, in either case, $x + x = 0$ for every x in F. Therefore if $x = y'$, where $y \in G$, then $y^2 = 1$, and this implies that $y = \pm 1 = 1$. Thus G has only one element, so that F has only two elements.

(This result shows that a 'logarithmic function' cannot be defined on the whole of the multiplicative group of a field, with values in the field, except in the trivial case of a field with only two elements. See **32**.)

62. (The property of F described by any of the three equivalent statements is called *completeness*. The system of real numbers is a complete totally ordered field, and can be shown to be essentially the only one: any two such fields are related by an order-preserving isomorphism, and are thus 'abstractly identical'; see **12, 13, 67**. For another characterization of completeness see **132**.)

63. If $r = \sqrt{p} + \sqrt{q}$ then
$$q = (r - \sqrt{p})^2 = r^2 + p - 2r\sqrt{p},$$
so that, since $r \neq 0$ (being a sum of positive numbers),
$$\sqrt{p} = (r^2 + p - q)/2r,$$
which is rational if r is rational.

64. For the first pair of inequalities, consider the identity
$$x^n - 1 = (x-1)(x^{n-1} + \dots + 1)$$
(which is related to the standard formula for the sum of n consecutive terms of a geometrical progression). The first inequality of the second pair shows that if $a > b > 0$ then $a^n > b^n$: thus distinct positive numbers have distinct nth powers.

65. Since $x - x = 0$, E contains 0. Hence if $x \in E$ then $-x \in E$, since $-x = 0 - x$. Hence if $x \in E$ and $y \in E$ then $x + y \in E$, since $x + y = x - (-y)$. Thus E is a subgroup of the additive group of the real numbers.

If there are infinitely many points of E between 0 and 1, consider a division of the interval $]0, 1[$ into k subintervals of equal length: at least one of these subintervals must contain infinitely many points of E (by 'the pigeon-hole principle'). Thus there is an interval of length $1/k$ which contains distinct points x and y of E: let $h = x - y$. Then $h \in E$ and, because E is an additive group, $nh \in E$ for every integer n. Also
$$0 < |h| < 1/k,$$
and so every interval of length $1/k$ contains nh for some integer n.

Now if λ is irrational and E consists of the numbers $m+n\lambda$, for all integers m and n, then distinct pairs of integers give distinct points of E (if $m+n\lambda = m'+n'\lambda$ then $m = m'$ and $n = n'$), and for each non-zero integer n there is an integer m such that

$$0 < m+n\lambda < 1;$$

hence there are infinitely many points of E between 0 and 1.

66. It is clear that c is an upper bound of E. Let d be a real number less than c. If n is any integer greater than $1/c$, there is a positive integer m such that

$$\frac{m^2}{n^2} < c \leqslant \frac{(m+1)^2}{n^2}.$$

Then

$$c - \frac{m^2}{n^2} \leqslant \frac{(m+1)^2}{n^2} - \frac{m^2}{n^2} = \frac{2m+1}{n^2} \leqslant \frac{3m}{n^2},$$

and

$$\left(\frac{3m}{n^2}\right)^2 = \frac{9}{n^2} \cdot \frac{m^2}{n^2} < \frac{9c}{n^2},$$

so that if n is large enough to ensure that $n^2 > 9c/(c-d)^2$ then

$$c - \frac{m^2}{n^2} < c-d,$$

and so

$$d < \frac{m^2}{n^2} < c.$$

Thus d is not an upper bound of E. Therefore $c = \sup E$.

(Constructions of this kind are often used in analysis. In devising them, one does not necessarily foresee all the details in advance: in this case, for instance, one begins with the idea of finding a rational number whose square is less than c but 'not too much less', and then makes the details precise as one goes along. See **14**.)

67. (A totally ordered field having the property described by any of the equivalent statements (i)–(v) is said to be *Archimedean*—though an idea resembling that expressed by (iv) was propounded by Eudoxus, a Greek mathematician who preceded Archimedes. Any subfield of an Archimedean field is itself Archimedean. A totally ordered field is necessarily Archimedean if it is complete. That the system of real numbers is essentially the only complete totally ordered field can be proved by considering (v); see **13**. Any Archimedean field can be isomorphically embedded—for example by Dedekind's construction†—in a complete

† J. W. R. Dedekind, 1831–1916; German.

totally ordered field; thus the system of real numbers is essentially the largest Archimedean field. Totally ordered fields that are not Archimedean are known to exist, but not within the present realm of elementary mathematics.)

68. Any subfield of the field of real numbers must contain 1, and hence all rational numbers. Hence any subfield containing $\sqrt{p}$ must contain F_x, which is easily shown to be a subfield. (See **12**.)

In any isomorphism between F_p and F_q, 1 must correspond to 1, and hence each rational number corresponds to itself. If, in such an isomorphism, $r + s\sqrt{p}$ is the element of F_p that corresponds to the element $\sqrt{q}$ of F_q, then $r^2 + s^2p + 2rs\sqrt{p}$ corresponds to q, so $rs\sqrt{p}$ must be rational; therefore if $\sqrt{p}$ is irrational then $r = 0$ or $s = 0$. If $r = 0$ then s^2p corresponds to q, so $s^2p = q$, which is impossible if p and q are distinct primes; if $s = 0$ then $r = \sqrt{q}$, which is impossible if q is a prime.

(This exercise shows that the real field has infinitely many nonisomorphic subfields. If $\sqrt{p}$ is rational then F_p is simply the rational field; if not, the irrational elements of F_p are called *quadratic surds* over the rational field, and two such surds are said to be *conjugate* to each other if their sum—and hence also their product—is rational. The idea of a quadratic surd is prominent in the general theory of 'algebraic fields', although that theory is mainly concerned with roots of polynomial equations of degree greater than 2. See **19, 20, 21, 71**.)

69. The recurrence relation determines u_{n+2} once u_n and u_{n+1} are known; and it shows that u_{n+2} belongs to any field that contains a, b, c, u_n, u_{n+1}. Hence, by induction, u_n is determined for every n if u_1 and u_2 are known, and u_n belongs to any field that contains a, b, c, u_1, u_2. In particular, u_n is real for every n if a, b, c, u_1, u_2 are real; but in such a case λ and μ may not be real, for example if a and c are positive and $b = 0$.

70. That C_1 is convex can easily be deduced from the triangle inequality (the modulus of the sum of two complex numbers does not exceed the sum of their moduli). The complement of C_2 is convex, so $C_1 \setminus C_2$ is the intersection of two convex sets. That $C_2 \setminus C_1$ is not convex can be seen by letting $z_1 = 3(1 - i)$ and $z_2 = -z_1$: in this case $z_1 \in C_2 \setminus C_1$ and $z_2 \in C_2 \setminus C_1$, but $\frac{1}{2}z_1 + \frac{1}{2}z_2 \in C_1$.

(A diagram can be helpful here, but of course it cannot be a substitute for a proof. Note that the idea of convexity is available for sets in 'n-dimensional Cartesian space'.)

71. Let $h = \zeta' - \zeta$, where ζ is an element of F such that $\zeta' \neq \zeta$ and $\zeta'' = \zeta$. Then $h \neq 0$, and

$$h' = \zeta'' - \zeta' = \zeta - \zeta' = -h;$$

so $h' \neq h$, and therefore $h \notin E$. But

$$(h^2)' = h'h' = (-h)^2 = h^2,$$

so $h^2 \in E$. Now $1 \neq -1$ in F, since $h \neq h' = -h$; hence division by 2 is possible in F. Since E is a field, $2^{-1} \in E$. If $z \in F$ and $z'' = z$, let

$$x = 2^{-1}(z + z'), \quad y = (2h)^{-1}(z - z').$$

Then $z = x + hy$, $x = x'$, and, since $(h^{-1})' = (h')^{-1} = -h^{-1}$, $y = y'$.

(It is shown here that if a field F has a non-trivial automorphism then it has a largest subfield E on which the automorphism acts trivially, and that the automorphism is inverse to itself—is an *involution*—if and only if F is isomorphic with a field obtained from E by the standard process of 'adjoining a square root'; $F \setminus E$ then consists of quadratic surds over E, conjugate surds corresponding to each other in the automorphism (see **19, 20, 21, 68**). The rational field has no non-trivial automorphism, since it has no proper subfield (see **11**). The automorphisms of a field form a group in an obvious way; in the branch of algebra known as 'Galois theory'—after the nineteenth-century French mathematician Évariste Galois—the solubility of polynomial equations within certain fields is related to structural properties of groups of automorphisms and associated groups of permutations. The insolubility by algebraic methods of the general equation of degree greater than 4 is established by this theory. See **19, 85**.)

72. The main point here is that the *squared length* of a vector $\mathbf{r}$ is, by definition, the scalar product $\mathbf{r}.\mathbf{r}$, and is positive unless $\mathbf{r} = \mathbf{0}$. Thus if $\mathbf{r} = \alpha_1 \mathbf{r}_1 - \alpha_2 \mathbf{r}_2$ then

$$0 \leqslant |\mathbf{r}|^2 = \alpha_1^2 |\mathbf{r}_1|^2 - 2\alpha_1 \alpha_2 \mathbf{r}_1.\mathbf{r}_2 + \alpha_2^2 |\mathbf{r}_2|^2,$$

equality holding on the left if and only if $\alpha_1 \mathbf{r}_1 = \alpha_2 \mathbf{r}_2$. So if $\mathbf{r}_1$ and $\mathbf{r}_2$ are such that $\alpha_1 \mathbf{r}_1 \neq \alpha_2 \mathbf{r}_2$ unless $\alpha_1 = \alpha_2 = 0$ we can show (by taking λ to be α_1/α_2 or α_2/α_1) that

$$|\mathbf{r}_1.\mathbf{r}_2| < |\mathbf{r}_1|\,|\mathbf{r}_2|.$$

(We have here a special case of the Cauchy–Schwarz–Bunjakowski inequality; see **18**. One often writes $\mathbf{r}^2$ for $|\mathbf{r}|^2$.)

73. (Here, if we write $f(\mathbf{r})$ for $\mathbf{r}'$, we have a mapping f, of the set of all 3-dimensional real vectors into itself, which 'preserves lengths and distances'. The argument shows that such a mapping 'preserves scalar products' and is 'linear' (see **131**). In fact it is easy to show that if f satisfies any two of the following conditions then it satisfies the third:

(i) f preserves length;

(ii) f preserves distances;

(iii) f is linear.

Simple examples show that a mapping f can satisfy any one of these conditions and fail to satisfy the others. A mapping that preserves distances is called an *isometry*, and is necessarily one-to-one; a linear isometry is sometimes called a *unitary transformation*. If g is any isometry, and f is defined by the equation

$$f(\mathbf{r}) = g(\mathbf{r}) - g(\mathbf{0}),$$

then f preserves lengths and distances, and $g(\mathbf{r}) = f(\mathbf{r}) + g(\mathbf{0})$ for every vector $\mathbf{r}$; thus every isometry is a composite of two isometries, of which one is a *translation*—addition of a fixed vector—and the other is length-preserving and consequently linear. These ideas and results are available for a large class of vector spaces with lengths and distances defined in terms of scalar products, and they have applications in several branches of mathematics. For an isometry of the Euclidean plane see **21**.)

74. By taking $\mathbf{r}_3$ to be $\mathbf{r}_1$ (or $\mathbf{r}_2$) in the identity

$$(\mathbf{r}_1 \wedge \mathbf{r}_2) \wedge \mathbf{r}_3 = \mathbf{r}_3 . \mathbf{r}_1 \mathbf{r}_2 - \mathbf{r}_2 . \mathbf{r}_3 \mathbf{r}_1$$

we can show that if $\mathbf{r}_1$ and $\mathbf{r}_2$ are linearly independent then $\mathbf{r}_1 \wedge \mathbf{r}_2 \neq \mathbf{0}$; the converse proposition follows from the fact that $\mathbf{r} \wedge (\alpha \mathbf{r}) = \mathbf{0}$. (The arguments here apply to 3-dimensional vectors over any field. See **24**.)

(Identities like

$$(\mathbf{r}_1 \wedge \mathbf{r}_2) \wedge \mathbf{r}_3 + (\mathbf{r}_2 \wedge \mathbf{r}_3) \wedge \mathbf{r}_1 + (\mathbf{r}_3 \wedge \mathbf{r}_1) \wedge \mathbf{r}_2 = \mathbf{0}$$

occur in contexts other than vector algebra. The general setting is that of an Abelian group, written in the additive notation, with a binary 'multiplication' × which is distributive but not necessarily associative. Such a system—which is a ring if the multiplication is associative—may be such that, for any of its elements x, y, z,

$$(x \times y) \times z + (y \times z) \times x + (z \times x) \times y = 0;$$

this condition is known as *Jacobi's identity*—after the nineteenth-century German mathematician C. G. J. Jacobi. An example other than that of vector algebra can be constructed by taking an arbitrary ring and defining

$x \times y$ to be the *commutator* $xy - yx$, for any elements x and y of the ring. In this case, as in that of vector algebra, we also have the identity

$$x \times x = 0.$$

Systems that satisfy this together with Jacobi's identity are associated with the name of the nineteenth-century Norwegian mathematician Sophus Lie; for instance, the system of 3-dimensional real vectors, with scalar products disregarded, belongs to a class of systems known as 'Lie algebras'. The identity $x \times x = 0$ implies, because of the distributive laws, that the operation $\times$ is *anti-commutative* in that

$$x \times y = -y \times x$$

for all elements x and y. When this condition is satisfied Jacobi's identity has the equivalent form

$$x \times (y \times z) + y \times (z \times x) + z \times (x \times y) = 0.)$$

75. To show that $\mathbf{p}'.\mathbf{q} = -\mathbf{q}'.\mathbf{p}$, consider $(\mathbf{p}+\mathbf{q}).(\mathbf{p}+\mathbf{q})'$.

To show that $\mathbf{r}' = \boldsymbol{\omega} \wedge \mathbf{r}$, note that if $\mathbf{r} = x\mathbf{i} + y\mathbf{j} + z\mathbf{k}$ then

$$\mathbf{r}' = x\mathbf{i}' + y\mathbf{j}' + z\mathbf{k}',$$

and that $\mathbf{i}' = \mathbf{i}'.\mathbf{i}\,\mathbf{i} + \mathbf{i}'.\mathbf{j}\,\mathbf{j} + \mathbf{i}'.\mathbf{k}\,\mathbf{k} = \mathbf{i}'.\mathbf{j}\,\mathbf{j} - \mathbf{k}'.\mathbf{i}\,\mathbf{k}$,   etc.

If $\mathbf{u} \wedge \mathbf{r} = \boldsymbol{\omega} \wedge \mathbf{r}$ then $(\mathbf{u} - \boldsymbol{\omega}) \wedge \mathbf{r} = \mathbf{0}$; and if this holds for every vector $\mathbf{r}$ then $\mathbf{u} - \boldsymbol{\omega} = \mathbf{0}$ (see **74**).

(If we think of each point of 'space' as being occupied by a moving particle in such a way that the distance between any two particles remains constant, we have the idea of a—space-filling—'rigid body' in motion. If $\mathbf{r}$ represents the position at a particular instant of a typical particle relative to some reference particle P of the body, and if $f_t(\mathbf{r})$ represents this relative position after a lapse of time t, then f_t is a linear isometry; see **73**. By making some assumptions about the existence of an instantaneous relative velocity $\mathbf{r}'$ for the particle at the relative position $\mathbf{r}$ in these circumstances, we can infer that the mapping $\mathbf{r} \to \mathbf{r}'$ is linear; and, since the rate of change of $|\mathbf{r}|$ is zero, that $\mathbf{r}.\mathbf{r}' = 0$. We then see that the motion of the whole body, relative to P, is instantaneously represented, through the equation $\mathbf{r}' = \boldsymbol{\omega} \wedge \mathbf{r}$,

by the vector $\boldsymbol{\omega}$, which is uniquely determined once P has been chosen. If we consider another reference particle Q, whose position relative to P is represented by a vector $\mathbf{q}$, then the position and velocity of the typical particle relative to Q are respectively $\mathbf{r} - \mathbf{q}$ and $\mathbf{r}' - \mathbf{q}'$; and since $\mathbf{q}' = \boldsymbol{\omega} \wedge \mathbf{q}$,

$$\mathbf{r}' - \mathbf{q}' = \boldsymbol{\omega} \wedge (\mathbf{r} - \mathbf{q}).$$

Thus $\boldsymbol{\omega}$ does not depend on the choice of reference particle; it is called the instantaneous 'angular velocity', or 'spin', of the body. These results can be used to discuss the motion of a rigid body which does not fill the whole of space: one imagines a 'rigid extension' of the moving body to the whole of space, such an extension being uniquely possible if the body is not 1-dimensional.)

76. If $\mathbf{u} \wedge \mathbf{r} = \lambda \mathbf{u} - \mathbf{v}$ then

$$0 = \mathbf{u}.(\mathbf{u} \wedge \mathbf{r}) = \lambda \mathbf{u}^2 - \mathbf{u}.\mathbf{v},$$

so $\lambda = \mathbf{u}.\mathbf{v}/\mathbf{u}^2$. The given identity suggests a solution of the form

$$\mathbf{r} = \alpha \mathbf{u} \wedge \mathbf{v},$$

where α is some scalar; in fact

$$\mathbf{u} \wedge (\alpha \mathbf{u} \wedge \mathbf{v}) = \mathbf{v}.\mathbf{u}\alpha\mathbf{u} - \mathbf{u}.\alpha\mathbf{u}\mathbf{v} = \alpha(\mathbf{u}.\mathbf{v}\mathbf{u} - \mathbf{u}^2\mathbf{v}),$$

so if $\lambda = \mathbf{u}.\mathbf{v}/\mathbf{u}^2$ we get a solution on taking α to be $1/\mathbf{u}^2$. But if $\mathbf{r}$ is a solution then so is $\mathbf{r} + \mathbf{q}$ if (and only if) $\mathbf{q}$ is a scalar multiple of $\mathbf{u}$.

(If the equation $\mathbf{r}' = \boldsymbol{\omega} \wedge \mathbf{r}$ (see **75**) is taken to represent the instantaneous motion of a rigid body relative to one of its particles P which has instantaneous velocity $\mathbf{v}$ relative to points fixed in space, and if Q is a fixed point whose position relative to P is represented instantaneously by $\mathbf{p}$, then the instantaneous velocity of a typical particle of the body relative to Q is $\mathbf{r}' + \mathbf{v}$, the position of this particle relative to Q is represented instantaneously by $\mathbf{r} - \mathbf{p}$, and

$$\mathbf{r}' + \mathbf{v} = \boldsymbol{\omega} \wedge (\mathbf{r} - \mathbf{p}) + \boldsymbol{\omega} \wedge \mathbf{p} + \mathbf{v}$$
$$= \boldsymbol{\omega} \wedge (\mathbf{r} - \mathbf{p}) + \lambda \boldsymbol{\omega}$$

if $\boldsymbol{\omega} \wedge \mathbf{p} = \lambda \boldsymbol{\omega} - \mathbf{v}$. Hence if we assume that $\boldsymbol{\omega} \neq \mathbf{0}$, and take λ to be $\boldsymbol{\omega}.\mathbf{v}/\boldsymbol{\omega}^2$, we see from the exercise—with $\mathbf{u} = \boldsymbol{\omega}$ and $\mathbf{r} = \mathbf{p}$—that Q can be chosen so that the instantaneous motion of the body relative to Q is compounded of a 'rotational motion' characterized by the angular velocity $\boldsymbol{\omega}$, and a 'translational motion' with velocity $\lambda \boldsymbol{\omega}$. Thus if $\lambda \neq 0$ the motion is instantaneously that of a 'screw'; λ is the 'pitch' of the screw, and the screw is 'right-handed' or 'left-handed' according as λ is positive or negative; Q is not uniquely determined but can be any point of a certain line, instantaneously fixed, whose direction is that of $\boldsymbol{\omega}$: this line is the 'axis' of the screw.)

77. If $A = \begin{pmatrix} 0 & 1 \\ 0 & 0 \end{pmatrix}$ and $B = \begin{pmatrix} 0 & 0 \\ 1 & 0 \end{pmatrix}$ then $AB \neq BA$; and $A^2 = O$, where $O = \begin{pmatrix} 0 & 0 \\ 0 & 0 \end{pmatrix}$.

(Sums of 2-by-2 matrices over F are defined by the rule

$$\begin{pmatrix} a_1 & b_1 \\ c_1 & d_1 \end{pmatrix} + \begin{pmatrix} a_2 & b_2 \\ c_2 & d_2 \end{pmatrix} = \begin{pmatrix} a_1+a_2 & b_1+b_2 \\ c_1+c_2 & d_1+d_2 \end{pmatrix}.$$

With this definition and that of matrix multiplication, the 2-by-2 matrices over F form a ring—in fact a non-commutative ring which cannot be isomorphically embedded in a division ring. This ring has a unit element, I, and it has a subring isomorphic with F. The matrices

$$\begin{pmatrix} x & y \\ -y & x \end{pmatrix}$$

form a subring which is isomorphic with the field of complex numbers if F is the field of real numbers, but is not a field if F is the field of complex numbers. The rules defining addition and multiplication of 2-by-2 matrices can be generalized in an obvious way to the case of n-by-n matrices over F, giving, for each natural number n, a ring with interesting properties; and F need not be a field, but can be any ring. The theory of matrices is a large branch of algebra, with many applications; it is closely connected with the theory of groups, with the theory of rings, and with 'linear algebra'—an important subject in which ideas of linearity are studied in a general setting based on the axiomatic concept of a vector space. In linear algebra one is often concerned with matrices in which the number of rows is not the same as the number of columns: for example, vectors can be regarded as matrices. The problem of *inverting* a matrix—finding its reciprocal, if this exists, in an appropriate ring—is important in certain kinds of numerical work; this problem is essentially that of solving a set of 'simultaneous' linear equations. See **131**.)

78. *The binomial theorem* states that if a and b are elements of a commutative ring, and n is any natural number, then

$$(a+b)^n = a^n + na^{n-1}b + \dots + \binom{n}{r} a^{n-r}b^r + \dots + b^n,$$

where, for $r = 0, 1, 2, \dots, n$, the coefficient $\binom{n}{r}$ is the natural number $\dfrac{n!}{r!(n-r)!}$ (the $(r+1)$th number in the $(n+1)$th row of 'Pascal's triangle'†).

† Blaise Pascal, 1623–1662; French.

If the ring is a field, the theorem can be stated equivalently in the form

$$(1+x)^n = 1+nx+ \ldots + \binom{n}{r} x^r + \ldots + x^n,$$

where x is any element of the field. If p is a prime number and r is a natural number less than p, then p is a factor of $\binom{p}{r}$ since p and $r!$ are factors of $p!/(p-r)!$ having no common factor.

Let p be any prime number, and let n be a natural number such that

$$n^p \equiv n \pmod{p};$$

for example, n could be 1. Then, by the binomial theorem applied to the ring of integers,
$$(1+n)^p = 1+np+ \ldots + n^p.$$

Since p is a factor of each term on the right other than the first and the last, it follows that

$$(1+n)^p \equiv 1+n^p \equiv 1+n \pmod{p}.$$

Hence, by induction, $n^p \equiv n \pmod{p}$ for every natural number n.

(The last result states that p is a factor of $n^p - n$. Since

$$n^p - n = n(n^{p-1} - 1),$$

it follows that p, being prime, is a factor of n or of $n^{p-1} - 1$. Thus, for any natural number n, if p is a prime number which is not a factor of n then

$$n^{p-1} \equiv 1 \pmod{p}.$$

This result is known as *Fermat's theorem*, after the seventeenth-century French mathematician P. Fermat; some interesting parts of the elementary theory of numbers are closely connected with it. See **184**.)

79. $\dfrac{n^2}{n^2-1} = 1 + \dfrac{1}{n^2-1}$, so, by the binomial theorem applied to the field of rational numbers,

$$\left(\frac{n^2}{n^2-1}\right)^n = 1 + \frac{n}{n^2-1} + \ldots > 1 + \frac{1}{n} + \ldots > 1 + \frac{1}{n},$$

the terms represented by ... being positive. The second of the required inequalities is equivalent to the first, as can be seen by writing it in the form

$$\left(\frac{n}{n-1}\right)^n > \left(\frac{n+1}{n}\right)^{n+1}$$

and noting that this is equivalent to

$$\left(\frac{n}{n-1}\right)^n\left(\frac{n}{n+1}\right)^n > \frac{n+1}{n}.$$

The third inequality is equivalent to

$$\left(1-\frac{1}{n^2}\right)^n < 1,$$

and we know that $x^n < 1$ if $0 \leqslant x < 1$.

The fourth inequality follows from the binomial theorem together with the fact that, for $r = 2, 3, \ldots, m+1$,

$$\frac{m(m-1)\ldots(m-r+1)}{m^r} = \left(1-\frac{1}{m}\right)\ldots\left(1-\frac{r-1}{m}\right)$$
$$< \left(1-\frac{1}{m+1}\right)\ldots\left(1-\frac{r-1}{m+1}\right)$$
$$= \frac{(m+1)m\ldots(m-r+2)}{(m+1)^r}.$$

The fifth inequality can be deduced from the third, with the aid of the fourth if $m < n$ or of the second if $m > n$.

(For each natural number n let $x_n = \left(1+\frac{1}{n}\right)^n$ and $y_n = \left(1-\frac{1}{n}\right)^{-n}$.

Then the facts established here include the following: the sequence $\{x_n\}$ is increasing, the sequence $\{y_n\}$ is decreasing, and each term of the first sequence is less than every term of the second. It follows that $\{x_n\}$ has a least upper bound x, that $\{y_n\}$ has a greatest lower bound y, and that $x \leqslant y$. In fact $x = y$, since $y - x < y_n - x_n$ and, as can be seen from **64**, $y_n - x_n < y_n/n$. The common bound is the irrational number e, which is encountered in the theory of logarithms.)

80. (The ring of characteristic functions of subsets of X is isomorphic to the ring formed, according to **8**, by the subsets themselves—the operation $\underset{2}{+}$ corresponds to $\triangle$, and multiplication in the ring of functions corresponds to $\cap$ in the ring of sets. It is an example of a Boolean ring. See **106**.)

81. To show that $\mathscr{A}$ is not a ring when A is the additive group of the integers, consider (for example) the functions f and g defined as follows:

$$f(a) = a+1, \quad g(a) = a \quad (a \in A).$$

Comparison of $f \circ (g+g)$ with $f \circ g + f \circ g$ in this case shows that $\mathscr{A}$, with $\circ$ as 'multiplication', is not subject to the left-hand distributive law.

(A function that belongs to $\mathscr{B}$ is known as an *endomorphism* of the Abelian group A; and $\mathscr{B}$ with + and o is *the ring of endomorphisms* of A. Such rings are important in the general theory of rings.)

82. By the second of the two given equations,

$$g(a) = g(a)f(0) - f(a)g(0)$$

whence it follows that $g(0) = 0$ and that if a can be chosen so that $g(a) \neq 0$ then $f(0) = 1$.

If $f(0) = 1$ the given equations show that, for any element b of A, $f(-b) = f(b)$ and $g(-b) = -g(b)$ (that is, that f is an *even* function and g is an *odd* function), and hence also that $|\chi(a)| = 1$ and $\chi(a+b) = \chi(a)\chi(b)$. With the further assumption about the range of f, we know that this range includes the interval $[-1, 1]$; so that if x and y are real numbers such that $x^2 + y^2 = 1$ then, since $-1 \leqslant x \leqslant 1$, there is an element a of A such that $f(a) = x$. But, by the first of the given equations,

$$\{f(a)\}^2 + \{g(a)\}^2 = f(0) = 1,$$

so if $f(a) = x$ then $g(a) = \pm y$, and hence $\chi(a)$ or $\chi(-a)$ is $x + iy$.

(This exercise isolates certain aspects of the theory of the circular functions—aspects which are essential to the theory of polar representation of complex numbers. It also illustrates the idea of a 'group character': a function χ that maps A into the set of complex numbers of unit modulus in such a way that $\chi(a+b) = \chi(a)\chi(b)$, for any elements a and b of A, is called a *character* of the Abelian group A; the characters of A form a group under pointwise multiplication, and this is called *the character group* of A. These ideas are fundamental to the subject known as 'harmonic analysis'. Note that a character χ has the property that $\chi(0) = 1$, and that if there is a non-zero element p of A such that $\chi(p) = 1$ then

$$\chi(a+p) = \chi(a)$$

for every element a of A, which is to say that χ is a *periodic* function, having p as a *period*; the periods, including 0, of any function on A form a subgroup of A. A complex-valued function on A has this property of periodicity if and only if its real and imaginary parts—the functions f and g in the case considered here—have the same property; and if it has the property then so does $\psi \circ \chi$ if ψ is any function defined on the unit circle—for instance if, when $|z| = 1$, $\psi(z)$ is a sum of constant multiples of z^n for integers n of which some may be negative. See **100**.)

83. If $x \geqslant 0$ then x is clearly an upper bound of the set $\{f(x') : x' \leqslant x\}$; and if $y < x$ there is a rational number x' such that $y < x' \leqslant x$ and consequently $y < f(x')$: thus x is the least upper bound, $\varphi(x)$. A similar argument shows that $\varphi(x) = 0$ if $x < 0$.

(Note that φ is continuous although f is not.)

84. If $x \in I$ then
$$f(x) + g(x) \leqslant \sup{(f+g)}(I),$$
and therefore, since $\inf f(I) \leqslant f(x)$,
$$\inf f(I) + g(x) \leqslant \sup{(f+g)}(I);$$
hence $\sup{(f+g)}(I) - \inf f(I)$ is an upper bound of $g(I)$, so that
$$\sup g(I) \leqslant \sup{(f+g)}(I) - \inf f(I).$$

85. (This exercise shows how a cubic equation of a certain type can be solved in terms of square roots and cube roots. The method gives what is known as 'Cardan's solution', after a sixteenth-century Italian mathematician who was given the idea by a contemporary, Tartaglia. Although it is of theoretical interest, it has little practical value: in numerical work, where rational approximations of prescribed accuracy are required, polynomial equations are usually 'solved' by systematic iterative methods, or by sequences of good guesses. See **35, 36, 37**.)

86. Let I be any bounded closed interval with more than one point. If $0 \in I$ then I contains $1/2n$ and $1/(2n+1)$ for some integer n, and therefore $g(I)$ contains 1 and -1, so that (from the definition of continuity) g is not continuous on I. If $0 \notin I$ then g is continuous on I since g restricted to I is then a composite of two continuous functions; hence in this case $g(I)$ is a bounded closed interval. If $0 \in I$ there is a bounded closed subinterval J of I which does not contain 0 but contains $1/2n$ and $1/(2n+1)$ for some integer n: then $g(J)$ is a bounded closed interval containing 1 and -1, so $g(I) \supseteq [-1, 1]$; but $g(I)$ is contained in the range of f, which, by hypothesis, is contained in $[-1, 1]$. Thus $g(I)$ is a bounded closed interval in every case.

(It is a fundamental property of any continuous function g of a real variable that if I is any bounded closed interval in the domain of g then $g(I)$ is a bounded closed interval. One might be tempted to conjecture that every function that has this property is continuous; but the exercise provides a counter-example, since, as is easily shown, a function f having the stated properties does exist.)

87. Consider (for example) the function f defined on $[1, 2]$ by the equation
$$f(x) = x^2 - 2.$$

88. From the identity $f(x+y) = f(x) + f(y)$ we deduce by induction that $f(nx) = nf(x)$ for every positive integer n and every real number x; in particular, $f(n) = nf(1)$. Also, $f(0) = 0$ since

$$f(0) + f(1) = f(0+1) = f(1).$$

If n is a negative integer then $-n$ is a positive integer and therefore

$$f(n) - nf(1) = f(n) + f(-n) = f(n-n) = 0.$$

Thus $f(n) = nf(1)$ for every integer n. If $r = m/n$, where m and n are integers and $n \neq 0$, then

$$nf(r) = f(nr) = f(m) = mf(1),$$

so $f(r) = rf(1) = ar$. If f is continuous, let x be any real number and let ϵ be any positive number. Then there is a positive number δ such that $|f(x) - f(r)| < \epsilon$ if r is any number such that $|x-r| < \delta$. But there is a rational number r such that $|x-r| < \delta$ and $|ax - ar| < \epsilon$: since $f(r) = ar$, it follows that $|f(x) - ax| < 2\epsilon$ and hence, since ϵ is an arbitrary positive number, that $f(x) = ax$.

(The last part of this argument has two ingredients which are used very often in analysis: we have appealed to the triangle inequality, and to the fact that 0 is the only number whose modulus is less than every positive number. See **126, 128**.)

89. To show that f is continuous, it is enough to show that f is continuous on $\{x: x \leq 1\}$ and also on $\{x: x \geq 1\}$; and on each of these intervals f is a composite of functions that are known to be continuous. (There is no need to appeal directly to a definition of continuity.)

90. By definition, $2^x = e^{x \log 2}$. Let y be any positive number. Since the domain of the logarithmic function contains y, and since $\log 2 \neq 0$, $f(x) = y$ if $x = \log y / \log 2$. Since the exponential function is continuous and monotonic, so also is f; hence f has a continuous monotonic inverse.

(The function f^{-1} gives logarithms to the base 2.)

91. From the definition

$$\log x = \int_1^x \frac{du}{u} \quad (x > 0),$$

and from fundamental properties of the integration process, it is easy to deduce that $\log x \leq x - 1$ (equality occurring if and only if $x = 1$). To

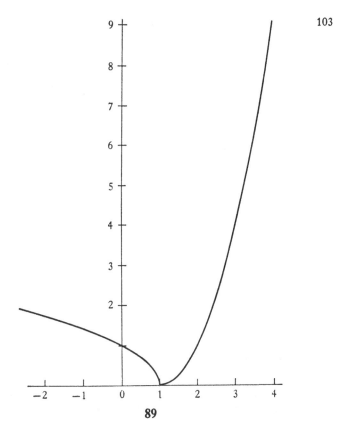

89

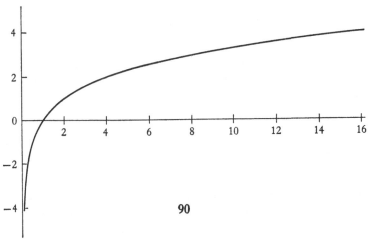

90

deduce the fourth of the required inequalities from the third, let

$$x_r = \frac{(p_1 + \dots + p_n) a_r}{p_1 a_1 + \dots + p_n a_n}, \quad \lambda_r = \frac{p_r}{p_1 + \dots + p_n},$$

for $r = 1, \dots, n$.

(The number
$$\frac{p_1 a_1 + \dots + p_n a_n}{p_1 + \dots + p_n}$$

is the *weighted arithmetic mean* of the numbers $a_1, \dots, a_n$, with *weights* $p_1, \dots, p_n$ respectively. The number

$$(a_1{}^{p_1} \dots a_n{}^{p_n})^{1/(p_1 + \dots + p_n)}$$

is the correspondingly-weighted *geometric* mean. The 'ordinary' means are obtained by taking $p_1 = \dots = p_n = 1$. Thus the exercise gives a generalization of the well-known inequality between the arithmetic and geometric means of two positive numbers. It is interesting to consider what is implied by the equality of the two means in the general case.)

92. The first proposition states that $f(x) \to 0$ as $x \to 0$; that is, that

$$\lim_{x \to 0} f(x) = 0.$$

(The concept of a limit is a subtle one, requiring careful thought for its proper understanding. To define it one has to make a statement whose logical structure involves a sequence of three quantifiers; and in arguments involving the concept it is easy to make serious logical mistakes—and hence serious mathematical mistakes—by taking the quantifiers in a wrong order.)

93. Let g be the function of a real variable defined by the equation

$$g(y) = |y|^\alpha.$$

Then g is continuous (being a composite of continuous functions), and $g(0) = 0$. Therefore if ϵ is any positive number there is a positive number η such that if $|y| < \eta$ then

$$g(y) = |g(y) - g(0)| < \epsilon;$$

but since $f(x) \to 0$ as $x \to 0$, there is a positive number δ (depending on η, and hence on ϵ) such that if $|x| < \delta$ then $|f(x)| < \eta$. Thus $g(f(x)) \to 0$ as $x \to 0$.

94. If $f(x) = 2^x$, for every real number x, then f is continuous, and $f(0) = 1$ (see **90**). Hence $f(x) \to 1$ as $x \to 0$. If $x < 0$ then $(b/a)^{1/x} \leqslant 1$, so that

$$a \leqslant (a^{1/x} + b^{1/x})^x = a\{1 + (b/a)^{1/x}\}^x \leqslant a 2^x;$$

hence $(a^{1/x} + b^{1/x})^x \to a$ as $x \to 0+$. If $x < 0$ then $(a/b)^{1/x} \leqslant 1$, so that

$$b \leqslant (a^{1/x} + b^{1/x})^x = b\{(a/b)^{1/x} + 1\}^x \leqslant b2^x.$$

95. (The function sinh is monotonic, and its range consists of all the real numbers; in fact—as one can see by solving a quadratic equation—if y is any real number then $\sinh x = y$ if $x = \log\{y + \sqrt{(y^2 + 1)}\}$, so that

$$\sinh^{-1} y = \log\{y + \sqrt{(y^2 + 1)}\}.$$

If, for any real number x, $u = \cosh x$ and $v = \sinh x$, then

$$u^2 - v^2 = 1;$$

and on the other hand, if u and v are any real numbers satisfying this equation there is a unique real number x such that $|u| = \cosh x$ and $v = \sinh x$. The set $\{(u, v): u^2 - v^2 = 1\}$ is an example of a 'hyperbola': hence the name 'hyperbolic functions'. The theory of these functions has some formal similarities with that of the circular functions, but is considerably simpler. See **100, 138, 139**.)

96. (If we restrict f to $\{x: x \geqslant 0\}$ or to $\{x: x \leqslant 0\}$ we get, in either case, a function differentiable at 0; the derivative at 0 being 1 in the first case and -1 in the second case.)

97. (This result, giving an 'expansion' for the nth derivative of a product of functions, is known as *Leibniz's theorem*, after the seventeenth-century German mathematician G. W. Leibniz, who shares with his English contemporary Isaac Newton the credit for inventing 'the differential calculus'.)

98. Let y be a real number. There is at most one (positive) prime number p such that $p^2 y$ is one of the numbers $p, p+1, \ldots, p^2 - 1$. If there is one such p, and $k = p^2 y$, then $y = f(x)$, where

$$x = \begin{cases} (k-1)/p^2 & \text{if} \quad k \neq p, \\ (p^2-1)/p^2 & \text{if} \quad k = p. \end{cases}$$

If there is no such p then $y = f(y)$. In either case there is exactly one real number x such that $y = f(x)$. Thus f^{-1} exists and has the same domain as f. Since $f(0) = 0$,

$$f'(0) = \lim_{x \to 0} \frac{f(x)}{x}$$

if this limit exists. Let δ be a positive number less than $\frac{1}{2}$, and suppose that $0 < |x| < \delta$. If $x = k/p^2$ then k cannot be $p^2 - 1$, and if k is one of the

5   W S E

numbers $p, p+1, ..., p^2-2$ then $f(x) = (k+1)/p^2$, so that

$$1 < \frac{f(x)}{x} = 1+\frac{1}{k} \leqslant 1+\frac{1}{p} \leqslant 1+x < 1+\delta.$$

Since $f(x)/x = 1$ in any other case, it follows that $f'(0) = 1$. Now if $0 < y < \delta$ and y is, for example, irrational, then $f^{-1}(y)/y = 1$; but if $y = 1/p$, where p is a prime number greater than $1/\delta$, then

$$f^{-1}(y) = (p^2-1)/p^2,$$

so that

$$\frac{f^{-1}(y)}{y} = p-\frac{1}{p} > p-1 \geqslant 2.$$

Thus f^{-1} is not differentiable at 0.

(This exercise gives a counter-example to the conjecture—which might be thought plausible—that if f is a function that maps the set of all real numbers one-to-one on to itself, and if f has a non-zero derivative $f'(x)$ at a point x, then f^{-1} is differentiable, with derivative $1/f'(x)$, at the point $f(x)$. This conjecture becomes a theorem if we add the hypothesis that f is continuous on some open interval containing x.)

99. When $x \neq 0$ the differential equation is equivalent to

$$f'(x)+x^{-2}f(x) = 0,$$

and this is equivalent to $\dfrac{\mathrm{d}}{\mathrm{d}x}\{e^{-1/x}f(x)\} = 0.$

(The 'integrating factor' $e^{-1/x}$ can be found by a standard procedure; see **138**.) It follows that, when $x > 0$,

$$f(x) = c\,e^{1/x},$$

where c is a constant; and since $f(1) = 1$, c must be e^{-1}.

(Note that the condition $f(1) = 1$ does not determine a solution of the differential equation outside the interval $\{x\colon x > 0\}$.)

100. The required identity can be established by induction or by 'the method of differences'. In either case one uses the identity

$$2\cos\varphi\sin\theta = \sin(\varphi+\theta)-\sin(\varphi-\theta).$$

Another method is based on the fact that if χ is the complex-valued function defined by the equation

$$\chi(\theta) = \cos\theta+i\sin\theta$$

then, for any integer k, $\cos k\theta$ is the real part of $\{\chi(\theta)\}^k$. Using this, we can deduce the required result from the polynomial identity

$$(1-z^2)(z+z^3+ ... +z^{2n-1}) = z(1-z^{2n})$$

(which is related to the standard formula for the sum of n consecutive terms of a geometrical progression). Thus, let $z = \chi(\theta)$: then $z^* = 1/z$, so that, if $z \neq z^*$,

$$z + z^3 + \ldots + z^{2n-1} = \frac{1 - z^{2n}}{z^* - z} = \frac{1 - z^{2n}}{-2i\sin\theta},$$

and now we have only to 'consider real parts'.

(The method of differences, for condensing an expression

$$a_1 + a_2 + \ldots + a_n,$$

depends on the discovery of a sequence $b_1, \ldots, b_n, b_{n+1}$ such that $a_k = b_{k+1} - b_k$ for $k = 1, 2, \ldots, n$; it can be used, for instance, when $a_k = 1/k(k+1)$—and one uses it, with the mean-value theorem, to prove the fundamental theorem of the calculus. The method involving the polynomial identity is more sophisticated, and has the advantage that if we consider imaginary parts instead of real parts we get an expression for

$$\sin\theta + \sin 3\theta + \ldots + \sin(2n-1)\theta;$$

in fact this method is a general one for deducing pairs of identities involving the circular functions from polynomial identities. In the expression

$$\cos\theta + \cos 3\theta + \ldots + \cos(2n-1)\theta,$$

and the corresponding expression with sin instead of cos, we have examples of 'Fourier series', so called after the nineteenth-century French mathematician J. B. Fourier; the general theory of such series is an important part of harmonic analysis. See **82**, noting that χ is a character of the additive group of the real numbers, and that its real and imaginary parts occur in the representation of 'simple harmonic motion' since they satisfy the differential equation of **137**. This function χ can be expressed in terms of the exponential function of a complex variable, through the formula

$$e^{i\theta} = \cos\theta + i\sin\theta,$$

which is associated with the name of the eighteenth-century Swiss mathematician Leonhard Euler; an adequate discussion of this requires an excursion into 'complex analysis'—a subject not understood in Euler's time; but one can easily see, with the aid of the addition formulae for cos and sin, that Euler's formula suggests a way to define complex powers of e—and hence of other numbers—which partly respects the 'laws of indices' and leads to a unified theory of the circular, exponential, and hyperbolic functions.)

APPENDIX 1

The examination papers from which Exercises 141–200 are taken relate to the following schedule (which was proposed in 1963 in connexion with 'the Swansea scheme' for pure mathematics in schools).

EXAMINATION FOR THE
GENERAL CERTIFICATE OF EDUCATION
OF THE
WELSH JOINT EDUCATION COMMITTEE

PURE MATHEMATICS

ADVANCED LEVEL (ALTERNATIVE SYLLABUS)

The examination will consist of two papers, and an optional Special paper, of 3 hours each. Questions may be set on any part of the syllabus in each of the papers, but a choice of questions will be allowed. The questions will not call for a high degree of ingenuity or for remarkable feats of memory, but candidates will be expected to show an understanding of the nature of mathematical reasoning, and to use symbols intelligently. No question will require the use of mathematical tables, or of any instrument other than a pen.

Syllabus

1. Sets (subsets of a given set): unions, intersections, complements; the empty set; de Morgan's laws; Cartesian products; one-to-one correspondence. Elements of mathematical logic.

2. Fundamental properties of the system of natural numbers (in particular, the commutative, associative, and distributive laws for addition and multiplication). The principle of mathematical induction, and simple applications of this. Scales of notation. Construction of the ring of integers (illustrating the ideas of equivalence class, Abelian group, isomorphism). Divisibility: expression of the highest common factor of m and n in the form $rm + sn$; Euclid's algorithm. Proof of the existence of infinitely many prime numbers; the sieve of Eratosthenes. Unique decomposition of a natural number into prime factors. Congruences.

3. Construction of the rational numbers. The concept of a field. Finite fields (in particular, the field of remainders modulo a prime). The concept of a totally ordered field. Moduli. Inequalities, and the laws governing their manipulation. Upper and lower bounds of sets in a totally ordered field; supremum and infimum. The non-existence of $\sqrt{2}$ in the field of rational numbers. Postulates for the system of real numbers. Existence of square roots of positive real numbers. Approximation of irrational numbers by rationals. Approximate computations and the checking of accuracy.

4. The field of complex numbers, and its identification with the Cartesian plane. Conjugates. Quadratic equations. The modulus of a complex number. Distance in the plane. The triangle inequality. The modulus of a product.

5. Algebra of 3-dimensional real vectors: linear combinations, linear dependence, the basis theorem; the scalar product and the vector product; determinants of order 3. Elements of the Euclidean geometry of lines and planes, treated algebraically.

6. The ring of polynomials in a single indeterminate, over any field: the remainder theorem; highest common factor; the field of quotients; partial fractions. The binomial theorem (for a positive integral index).

7. The idea of a function. Images and counter-images of sets. Restriction and extension of functions. Functions of a real variable: step-functions, rational functions, bounded functions, unbounded functions, monotonic functions; inverse functions; composite functions; linear combinations, products, quotients of functions. Sketching of graphs in simple cases. Continuity in the large (that is, uniform continuity on bounded closed intervals). Fundamental mapping properties of continuous functions. Continuity of combinations of continuous functions. Existence and continuity of simple algebraic functions.

8. Upper and lower Riemann sums of a bounded function on a bounded closed interval. The idea of integrability and the idea of area. Integrability of continuous functions and of bounded monotonic functions. Fundamental properties of the integration process (linearity, positivity, additivity). (Knowledge of Darboux's theorem on integrability will not be required.)

9. Definition and fundamental properties of the logarithmic function (of a positive real variable). The number e, and the exponential function. Meaning of a^x, where a is positive and x is real. Logarithms to bases other than e.

10. The idea of a limit. Continuity in terms of limits. Fundamental

theorems about limits of compound functions. Differentiability. Theorems underlying the technique of differentiation. Differentiation of rational functions, of the logarithmic function, of the exponential function, and of simple compounds of these.

11. The mean-value theorem and some of its applications, including 'the fundamental theorem of the calculus'. Principles of systematic integration. Calculation of logarithms. Linear differential equations of the first order.

12. Analytical definition of the number π and of the circular functions (of a real variable). Fundamental properties of these functions and of their inverses. Polar representation of complex numbers, and simple applications of this. The length of a simple differentiable arc in the plane; definition of angle. (No knowledge of infinite series will be required.)

Notes

(i) The order in which the topics are arranged in the above list is compatible with the logical structure of the subject, and could therefore be followed in a course of study. Some variations of this order could however be made without loss of coherence: for instance the item listed fifth could be taken at a later stage, since nothing else in the syllabus depends on it.

(ii) Items 1–9 represent a lower level of sophistication than the rest of the syllabus, in that they involve no explicit consideration of local properties of functions. These items, with the possible omission of the fifth, might form the basis of a first-year course. Anyone embarking on such a course should, if possible, have had a preview of the subject, and to this end some of the material of the first two or three items might be incorporated in an Ordinary-level course.

(iii) Items 7–12 call for some skill and imagination on the part of the teacher, and appreciable intellectual effort on the part of the pupil; but the difficulties should not be exaggerated. It is important to avoid unrigorous (that is, question-begging) arguments, but this is not to say that one should never appeal to a theorem that one has not proved. In elementary analysis there are several crucial theorems that are quite hard to prove, and in an introductory course it is legitimate to state these theorems without proof (provided that the individual pupil who wishes to study a proof is given an opportunity of doing so). Honest omission of certain proofs, or parts of proofs, is much less dangerous than the use of vague or incomplete definitions.

APPENDIX 2

AN APPROACH TO
ELEMENTARY ANALYSIS

(A note based on an address given by J. D. Weston at a symposium
held in April 1964 at Syracuse, Sicily, in celebration of the work of
Archimedes.)

Elementary analysis is essentially the theory of the calculus, and one of
its main themes is the possibility of inferring global properties of functions from local properties. (We make such an inference when, for
example, we obtain the general solution of a differential equation.) The
concept of a local property is a fairly subtle one; it was formed at a
comparatively late stage in the development of mathematics, and it was
not thoroughly understood until quite recent times. This, no doubt, is
one of the reasons for the widespread practice of teaching the calculus
without giving adequate attention to its logical foundations.

Many books have been written in which the calculus is presented as
though it were an inductive science like physics. This may sometimes
have been done through ignorance, but probably in most cases it has
been done in the fallacious belief that a difficult subject can be made
easier by concealing its logical structure from the student, at least in the
early stages of instruction. That this kind of approach is unsatisfactory
becomes increasingly clear as 'pre-calculus mathematics' becomes
increasingly pervaded by the spirit of pure mathematics. An intelligent
student who has tasted the delights of the axiomatic method will want
the calculus to be presented as a strictly deductive discipline.† It is still
not easy to do this, but some of the traditional difficulties can now be
reduced.

Although local properties are relatively difficult to understand, it is

† After describing the joy of his early introduction to 'Euclid', Bertrand Russell
wrote (in the first volume of his *Autobiography*, published in 1967) 'My
mathematical tutors had never shown me any reason to suppose the Calculus
anything but a tissue of fallacies.' The kind of teaching to which he had
presumably been exposed was denounced as 'an educational sham' and
'a sin against the spirit of mathematical progress' as long ago as 1889 by
Chrystal (in the preface to the second volume of his *Algebra*), but it has not
yet been generally abandoned.

usual to introduce them at or near the beginning of a course in elementary analysis, with little or no preliminary discussion of global properties. Generally, local continuity is discussed before uniform continuity, and the notion of a derivative is discussed before that of an integral. However, a reversal of this order can be advantageous, particularly (but not exclusively) in the case of students who, though not unintelligent, cannot give much time to the study of mathematics. Let us consider the main features of a course based on this pedagogical notion.

From the axiomatic point of view, an appropriate starting-point for a course in elementary analysis is the concept of a totally ordered field, of which the system of rational numbers is a familiar example. It is easy to show that the algebraic rules by which inequalities are manipulated are valid for any totally ordered field, and that the system of rational numbers is isomorphically embedded in any such field (and, indeed, that it is essentially the only minimal one). One can then introduce the idea of exact bounds of sets in a totally ordered field, and prove that if the field is *complete*, in the sense that every non-empty set that has upper bounds has a least upper bound, then every positive element has a square root in the field. This motivates the postulation of the system of real numbers as a complete totally ordered field. That such a system exists and is unique to within isomorphism should be stated, but need not be proved at this stage. (It is easy enough to define the real numbers in terms of the rational numbers, as Dedekind sections for example, but to verify that one then has a system with all the required properties is a somewhat tedious matter.)

The system of real numbers having been postulated in this way (or in some equivalent way), the idea of a function of a real variable can be considered, and illustrated by simple examples. Such functions can be classified according to certain global properties that they may or may not have, including boundedness, monotonicity, univalence, representability by rational expressions. Continuity also can be defined at this stage as a global property, in fact as uniform continuity on bounded closed intervals. The fundamental mapping theorems for continuous functions, and the usual theorems about the continuity of compound functions, can be deduced from this definition with no greater difficulty than one has with local continuity; indeed some simplifications are possible. Once these theorems are available it is very easy to establish the existence and continuity of various algebraic functions, and the student has a substantial amount of useful equipment at his disposal. (The fundamental mapping theorems state that if f is a continuous function on a bounded closed interval I then (i) the set $f(I)$ is a bounded closed interval, and (ii) if f is also univalent on I—which, by virtue of (i), is the case only if f is strictly monotonic on I—then the inverse function f^{-1} is continuous on $f(I)$.)

The next step is to consider upper and lower Riemann sums of a bounded function on a bounded closed interval, and to derive the notion of integrability. With the definition of continuity that has been adopted, it is trivially easy to prove that continuous functions are integrable. Some numerical computation of integrals can now be done; this serves to emphasize that integration is essentially a process of approximation, and that an approximation is worthless unless it is accompanied by an estimate of accuracy (obtained in this case by computing both upper and lower sums). The fundamental properties of the integration process— linearity, positivity, additivity, integrability of the modulus—can be established in a straightforward way, but in a short course these need only be stated. One is then in a position to define the logarithmic function and to establish its functional equation ($\log xy = \log x + \log y$), its continuity, and the fact that it has a continuous inverse, which is the exponential function. This provides an efficient way of assigning an unambiguous meaning to the symbol a^x, where $a > 0$ and x is any real number. The circular functions can be introduced in an essentially similar way (independently of trigonometry), but the detailed study of these functions is best deferred until the calculus has been more fully developed.†

It may be presumed that a student who has reached this stage in the study of analysis is prepared to appreciate the concept of a limit. Continuity can now be expressed in terms of this concept through Heine's theorem (which states that a function has the global property of being uniformly continuous on a bounded closed interval I if it has the local property of being continuous at each point of I), and the fundamental theorems about limits of compound functions can be proved by arguments of a type with which the student is now familiar. At this point the concept of a derivative can be introduced, and the fundamental theorems on which the technique of differentiation is based can be proved without difficulty. By way of the mean-value theorem one then reaches the fundamental theorem of the calculus, which confers the power of evaluating integrals by 'antidifferentiation'; the technique of this ('systematic integration') can be developed as part of the wider, and more interesting, technique of solving differential equations, and it is natural to develop concurrently the theory of the elementary transcendental

---

† An account of the circular functions which is suitable from this point of view is given in the pamphlet *Notes on the circular functions* (second edition, Swansea, 1966) by J. D. Weston. 14 pp.

functions of a real variable (the non-algebraic functions already mentioned, and functions simply related to these). It is by no means necessary to use infinite series to represent either numbers or functions in work at this level; it is only in more advanced analysis that infinite series are genuinely required, and their theory and usefulness can perhaps be best appreciated by the student who already has a sound knowledge of the calculus.

The introduction to analysis thus briefly described does not differ greatly from what is now usual in first-year university courses; but the order of development allows some economy in the use of time, and gives the student an opportunity to acquire some understanding of analysis while he is developing his skill in the calculus. It is also to some extent a historical order: Archimedes knew something about integration; a long time later came the idea of logarithms, and then the differential calculus. Of course it remains true that some of the fundamental theorems are rather hard to prove. The non-specialist student should learn to appreciate the significance of these theorems, but he need not be obliged to learn proofs of them.

APPENDIX 3

For ease of reference, some definitions and explanations not given elsewhere in the book are collected here. Certain terms denoting very primitive notions—for instance set, function (mapping), one-to-one correspondence, natural number, proof (deduction)—or very elementary concepts—for instance union, equivalence relation, integer (difference of natural numbers), rational number (quotient of integers), polynomial—are not explicitly defined anywhere in the book; but see the index for guidance as to their meanings and use.

For sets A and B (subsets of a given set), *the complement of B relative to A*, written $A \setminus B$, is $\{x: x \in A, x \notin B\}$; this is $\varnothing$ if $A \subseteq B$.

When it is understood that all sets under discussion are subsets of a certain given set X, the complement of a set E relative to X is referred to simply as *the complement of E*, and may be denoted by cE. Thus cE is the set H, uniquely determined by E, such that

$$E \cup H = X \quad \text{and} \quad E \cap H = \varnothing.$$

Hence $ccE = E$, for any subset E of X.

By a *family of sets* we mean an assignment of a subset E_ι of X to each member ι of some 'index set' I; that is, we mean a function whose domain is I and whose values are subsets of X. For example, I might consist of the numbers 1 and 2, in which case we might prefer to write, say, E and H rather than E_1 and E_2; but I might be an infinite set (as in the case of an infinite sequence of sets, which is a family whose index set consists of all the natural numbers). The sets E_ι need not be distinct (a function need not be univalent).

De Morgan's laws state that, for any family of sets, (i) the complement of the union of the sets is the intersection of their complements, and (ii) the complement of the intersection of the sets is the union of their complements: in symbols,

(i) $c \bigcup_{\iota \in I} E_\iota = \bigcap_{\iota \in I} cE_\iota$,   (ii) $c \bigcap_{\iota \in I} E_\iota = \bigcup_{\iota \in I} cE_\iota$,

or, in the special case of two sets E and H (possibly identical),

(i) $c(E \cup H) = cE \cap cH$,   (ii) $c(E \cap H) = cE \cup cH$.

Since each set is the complement of its complement, the laws (i) and (ii) are 'dual' to each other in that each can be deduced from the other by 'taking complements'. (Augustus De Morgan, 1806–1871; British, first president of the London Mathematical Society.)

The *Cartesian product*, $\prod_{\iota \in I} E_\iota$, of a family of sets is the set of all functions f that map I into $\bigcup_{\iota \in I} E_\iota$ in such a way that $f(\iota)$ is an element of E_ι for each ι in I. In the special case of two sets E and H, this product is often denoted by $E \times H$, and consists of all ordered pairs (x, y) for which $x \in E$ and $y \in H$. The *Cartesian plane* is $R \times R$, where R is the set of all real numbers; since $R \times R$ consists of the complex numbers, it is often called *the complex plane*; and when considered with the usual notion of distance it is called *the Euclidean plane*. (René Descartes, or Des Cartes, 1596–1650; French.)

If F is a field, a *vector space over F* is an Abelian group G, written in the additive notation, together with a mapping $(\alpha, x) \to \alpha x$ of $F \times G$ into G such that

(i) $\alpha(x+y) = \alpha x + \alpha y$; (ii) $(\alpha+\beta)x = \alpha x + \beta x$;
(iii) $(\alpha\beta)x = \alpha(\beta x)$ $(=\alpha\beta x)$; (iv) $1x = x$,

for any elements α, β of F and any elements x, y of G. Linear combinations (weighted sums) of elements of a vector space are defined in an obvious way (as for 3-dimensional real vectors). A set in a vector space is said to be *linearly independent* if no element of it is a linear combination of other elements of it. A *basis* of a vector space is a linearly independent set B such that every element of the space is a linear combination of elements of B. In a vector space over F, if $x_1 \neq x_2$ the *line* through x_1 and x_2 is

$$\{x: x = \alpha x_1 + (1-\alpha) x_2, \alpha \in F\}.$$

(If, in the definition of a vector space, we relax the condition that F be a field, stipulating only that it be a ring, and not insisting on (iv), we obtain the definition of a *module* over a ring—a *unital module* if the ring has a unit element and (iv) holds. Any Abelian group can be regarded as a unital module over the ring of integers; and every ring is a module over itself. Thus the theory of modules includes the theory of Abelian groups, the theory of rings, and the theory of vector spaces.)

An *interval* is a non-empty set I of real numbers such that if $x_1 \in I$ and $x_2 \in I$ then every number between x_1 and x_2 belongs to I. (In other words, an interval is a convex set of real numbers.) An *end-point* of an interval I is inf I (left-hand) or sup I (right-hand). The end-points of I constitute a set E_I which may have 0, 1, or 2 points; and I is said to be *closed* if $E_I \subseteq I$, *open* if $E_I \cap I = \varnothing$. If $a < b$ there are four intervals whose end-points are a and b:

$$[a,b] = \{x: a \leqslant x \leqslant b\}; \quad]a,b[= \{x: a < x < b\};$$

$$[a,b[= \{x: a \leqslant x < b\}; \quad]a,b] = \{x: a < x \leqslant b\}.$$

The *length* of each of these intervals is $b - a$. The length of an interval consisting of one point is defined to be 0. A *compact* interval is an interval that is bounded and closed. A *subinterval* of an interval I is an interval contained in I.

A real-valued function f whose domain is an interval I is called a *step-function* if (i) for each compact subinterval J of I, $f(J)$ is a finite set, and (ii) for each value y of f, $f^{-1}(\{y\})$ is a union of finitely many intervals. For a step-function f with bounded domain, each value y determines finitely many disjoint bounded intervals, no two of which have a common end-point, whose union is $f^{-1}(\{y\})$; the sum of their lengths, λ_y say, is uniquely determined by f and y, and we denote the sum of all the products $y\lambda_y$ by $\int f$. (If f is a non-negative step-function with bounded domain, $\int f$ is the area of a certain rectilinear figure in the Euclidean plane.)

If f is a real-valued or complex-valued function of a real variable, to say that f is *continuous on a compact interval* J in the domain of f is to say that for each positive number ϵ a positive number δ can be found with the property that if $x \in J$ and $x' \in J$ and $|x' - x| < \delta$ then $|f(x') - f(x)| < \epsilon$. If I is any interval in the domain of f, to say that f is *continuous on I* is to say that f is continuous on every compact subinterval of I. (For a function f whose domain is a non-degenerate interval I, it is clear that if f is continuous on I then, for any point x of I, f is *continuous at x* in the sense that $f(x') \to f(x)$ as $x' \to x$. Heine's theorem† asserts that a function is continuous on an interval I if it is continuous at each point of I.) Composites of continuous functions are continuous, and so are functions obtained from continuous functions by rational processes (addition, subtraction, multiplication, and division); hence, for example, every rational function is continuous on each interval contained in its domain. (Important properties of continuous functions are expressed by 'the fundamental mapping theorems': see Appendix 2.)

† H. E. Heine, 1821–1881; German.

Let I be a bounded interval, let f be a real-valued function whose domain includes I, and suppose that f is *bounded on* I (that is, that the image $f(I)$ is a bounded set). If g and h are step-functions with domain I, and if

$$g(x) \leqslant f(x) \leqslant h(x)$$

for every x in I, then $\int g$ is a *lower Riemann sum*, and $\int h$ an *upper Riemann sum*, for f on I. Let L be the set of all the lower Riemann sums, and U the set of all the upper Riemann sums, for f on I. These sets are non-empty, since I and $f(I)$ are bounded; moreover, each point of L is a lower bound of U, and each point of U is upper bound of L. Hence $\sup L \leqslant \inf U$. The function f may be such that $\sup L = \inf U$; f is then said to be *integrable* on I, in the sense of Riemann, and the common bound is the Riemann *integral* of f over I, written $\int_I f$. If a and b are the end-points of I, and $a \leqslant b$, this integral is also written as $\int_a^b f(x)\,\mathrm{d}x$ (without ambiguity, since the integral is unaltered if I is altered by the inclusion or omission of an end-point); and, by definition, $\int_b^a f(x)\,\mathrm{d}x = -\int_I f$.

A function is integrable on a compact interval I if it is continuous on I, or if it is monotonic on I, but not if it is unbounded on I. A function integrable on an interval I is integrable on every subinterval of I. The integration process has the following properties:

Linearity. If f and g are integrable on I, and if α and β are real constants, then $\alpha f + \beta g$ is integrable on I, and

$$\int_I (\alpha f + \beta g) = \alpha \int_I f + \beta \int_I g.$$

(This is true for complex constants α and β if integrals of complex-valued functions are defined in the obvious way, in terms of real and imaginary parts.)

Positivity. If f is integrable on I and $f(x) \geqslant 0$ for every x in I then $\int_I f \geqslant 0$; and (*strict positivity*) if f is continuous on I and $f(x) \geqslant 0$ for every x in I then $\int_I f = 0$ only if I has only one point or $f(x) = 0$ for every x in I.

Additivity. If f is integrable on I and on J, where I and J are disjoint intervals such that $I \cup J$ is an interval, then f is integrable on $I \cup J$, and

$$\int_{I \cup J} f = \int_I f + \int_J f.$$

If f is integrable on I then so is $|f|$, and

$$\left|\int_I f\right| \leqslant \int_I |f|;$$

and if $c \in I$ and F_c is defined on I by the formula

$$F_c(x) = \int_c^x f(u)\, du \quad (x \in I)$$

then F_c is continuous on I.

(G. F. B. Riemann, 1826–1866; German.)

Let x be a point in the domain of a function f (real-valued or complex-valued), and suppose that there is an open interval I containing x such that the domain of f either contains I or else contains all points of I on one side of x and no points of I on the other side. Then there may be a number $f'(x)$ such that

$$\frac{f(x')-f(x)}{x'-x} \to f'(x) \quad \text{as} \quad x' \to x;$$

if so, f is said to be *differentiable at* x, and the number $f'(x)$, which is uniquely determined by f and x, is called *the derivative of f at x*; it is often denoted by $Df(x)$ or by

$$\frac{d}{dx}f(x).$$

The technique of differentiation (calculating derivatives) is based on this definition and on the following five propositions deducible from it:

(1) If $f+g$ and fg are defined in the usual way (pointwise) for functions f and g which are differentiable at x then

$$(f+g)'(x) = f'(x)+g'(x),$$
$$(fg)'(x) = f(x)g'(x)+f'(x)g(x).$$

(In particular, differentiation is a linear process.)

(2) If f is differentiable at x and $f(x) \neq 0$ then

$$(1/f)'(x) = -f'(x)/\{f(x)\}^2.$$

(3) A composite function $f \circ g$ is differentiable at x if f and g are differentiable at $g(x)$ and at x respectively, and in this case

$$(f \circ g)'(x) = f'(g(x))g'(x).$$

(4) If the domain of f is an interval on which f is continuous and strictly monotonic, if f is differentiable at x, and if $f'(x) \neq 0$, then the inverse function f^{-1} is differentiable at $f(x)$, and

$$(f^{-1})'(f(x)) = 1/f'(x).$$

(5) If f is continuous on an interval I having more than one point, and if

$$F_c(x) = \int_c^x f(u)\,du \quad (x \in I),$$

then F_c is differentiable at each point x of I, and

$$F_c'(x) = f(x).$$

Differentiability of vector-valued functions of a real variable—an important idea in kinematics and in differential geometry—can be defined in an obvious way. Rules for differentiating scalar products and vector products can be easily deduced from the rule for products of real-valued functions.

The fundamental theorem of the calculus states that if f is integrable on $[a, b]$, and if F is a function which is continuous on $[a, b]$ and is such that $F'(x) = f(x)$ for every point x of $]a, b[$, then

$$\int_a^b f(x)\,dx = F(b) - F(a).$$

The number e is $\log^{-1} 1$; that is, e is defined by the equation

$$\int_1^e \frac{1}{x}\,dx = 1.$$

Since $\quad 1 = \log' 1 = \lim_{x \to 0} \frac{\log(1+x)}{x} = \lim_{x \to 0} \log(1+x)^{1/x},$

and since $\log^{-1}$ is a continuous function,

$$e = \lim_{x \to 0} (1+x)^{1/x}.$$

The number π, encountered in the theory of the circular functions, is defined by the equation

$$\pi = 4 \int_0^1 \frac{1}{1+x^2}\,dx.$$

For analytical definitions of the circular functions, and deductions of their fundamental properties, the reader is referred to the pamphlet mentioned in the footnote on p. 113. This pamphlet has the following section-headings:

The number π, and the functions sin *and* cos;
The addition formulae;
The functions tan, sec, cot, cosec, *and* $\tan^{-1}$;
The irrationality of π^2;
The integration of rational functions;
The functions $\sin^{-1}$ *and* $\cos^{-1}$;
Polar forms;
Arc-length and angle;
Approximation to $\sin\theta$ *and* $\cos\theta$ *by polynomials in* θ;
The circular functions of a complex variable.

INDEX

The numbers refer to pages. Reference is made to definitions of terms (on pages indicated here by *italic numerals*), to topics on which there are exercises, to important ideas that are discussed or mentioned, and to words whose mathematical uses are exemplified or explained. (Certain 'ordinary' English words have special meanings in mathematical contexts, and the correspondence between the words and their meanings is not one-to-one.) Some of the cross-references indicate significant linkages of ideas: the reader may find it instructive to consider the nature of these linkages, and to trace others.

126 INDEX

determine 2, 46, 106; *and see* calculation, establish, uniquely determined
determinant 27, 32, 42, 67, 82, 109
development of mathematics 56, 111, 114
diagram 78, 92; *and see* graph of a function
difference 47, 115; *and see* method, minus, subtraction, symmetric
differentiability 9, 21, 43–4, 49, 106, 109, *119*, 120
differential calculus 82, 105, 114; *and see* differentiation
equation 21–2, 34–7, 39, 40, 44, 46–7, 106–7, 110–11, 113
geometry 120; *and see* arc-length
differentiation 9, 10, 21, 39, 41, 46–7, 110, 113, 119–20
dimension 66
direction of a line 28, *29*, 45; *and see* axis, orthogonal
of a non-zero vector *29*
disjoint sets 1, 11, 51, 56, 117–18
disprove 52; *and see* counterexample, negation
dissecting 41; *and see* division, disjoint
distance 15, 27–9, 94–5, 109; *and see* close to, isometry, length, modulus, triangle inequality, radius
distinction 59
distributive law 43, 48, 56–7, *89*, 94–5, 99, 108
divisibility 17, 38, 40, 47, 108; *and see* congruence, factor, prime, remainder
division 93, 117; *and see* Euclid's algorithm, factorization, quotient, reciprocal, subinterval
by 2 93
ring *69*, 70, 97; *and see* field
divisors of zero 30, *55*, 89, 97; *and see* division ring, matrix
domain of a function 6, 20–1, 33, 41, 115, 117–19; *and see* function, non-empty set
dynamical systems 37

e 22, 44, 99, 107, 109, *120*; *and see* exponential function
educational sham 111
eighteenth century 24, 63, 107

electrical engineers 63
elementary algebra 57
analysis 110–14
arithmetic 73
transcendental functions 113–14
element of a set 1, 11, 44; *and see* belongs to, point, subset
elements of mathematical logic 108; *and see* reasoning
elimination 82
embedding 12, 23–5, 29, 63, 91, 97, 112; *and see* contain, identification, isomorphism
empty set (∅) 1, 6, 38, 59, 108, 115
end-point 9, 78, 83–4, *117*, 118
endomorphism of an Abelian group *100*
of a vector space *33*
equal (=) 51
equality 15, 28, 102, 104; *and see* equation, identity, inclusion, inequality
equation 48; *and see* cubic, differential, identity, linear, polynomial, quadratic, quartic, simultaneous
equipment 112
equivalence class 24, 29, 40, 42, 108; *and see* coset, direction, integer, point (of a projective plane), rational number
relation 11, 24, 29, 40, 115; *and see* congruence, equality
equivalent statements 13–14, 23, 27, 32; *and see* implication
Eratosthenes *see* sieve
error 44, 52, 75, 81, 86; *and see* accuracy, computation, reasoning
establish 51, 58; *and see* prove
estimate 3; *and see* accuracy, approximation, inequality
Euclid 111
Euclidean geometry 29, 109
plane 94, *116*, 117
Euclid's algorithm 40, 53, 70, 108
Eudoxus 91
Euler, Leonhard 107
Euler's formula *107*
evaluate *see* calculation
even function *100*; *and see* cosh, cosine, even power
number 1, 25, 49, 89
power 43